SURFACE ACTIVITY
OF
PROTEINS

Chemical and
Physicochemical Modifications

edited by
Shlomo Magdassi

The Hebrew University of Jerusalem
Jerusalem, Israel

Marcel Dekker, Inc.   New York • Basel • Hong Kong

Library of Congress Cataloging-in-Publication Data

Surface activity of proteins : chemical and physicochemical
modifications / edited by Shlomo Magdassi.
  p. cm.
  Includes index.
  ISBN 0-8247-9532-6 (hardcover : alk. paper)
  1. Protein engineering. 2. Proteins—Surfaces. 3. Proteins—
Chemical modification. I. Magdassi, Shlomo.
  TP248.65.P76S87 1996
  660'.63—dc20   96-25981
                                                                  CIP

The publisher offers discounts on this book when ordered in bulk
quantities. For more information, write to Special Sales/Professional
Marketing at the address below.

This book is printed on acid-free paper.

Marcel Dekker, Inc.
270 Madison Avenue, New York, New York 10016

Current printing (last digit):
10 9 8 7 6 5 4 3 2 1

PRINTED IN THE UNITED STATES OF AMERICA

Preface

Proteins are biopolymers that are encountered in many applications, such as food emulsions, hair conditioners, photographic emulsions, and various medical diagnostic products. Many of these applications are frequently based on the unique surface activity of the proteins, which is reflected in functional properties such as foaming, emulsification, and gelling. The proteins are composed of polymeric chains containing many hydrophobic and hydrophilic domains, often giving the molecules an amphipathic structure somewhat similar to that of polymeric surfactants.

Because the functional properties of the proteins are strongly dependent on their molecular characteristics this should be taken into account in order to select the proper protein for an application.

In general, using a protein in a given application requires a combination of several properties, such as gelling with emulsification, adsorption with steric stabilization, and so forth. Improved functional properties can often result from enhancement of the surface activity of the protein. Alteration of surface activity of proteins can be achieved by various methods, and among them chemical modifications seem to be the most promising. This book deals with the modification of the surface activity of proteins from two points of view: first a description of specific modification methods, such as increasing the negative charges by succinylation, or attachment of surfactant molecules without formation of covalent bonds, and second evaluation of the implication of the various modifications on the surface activity, using examples of specific systems and applications.

The book gathers several approaches to modifying the surface activity of proteins, and therefore it contains both theoretical and practical descriptions. After an introductory chapter describing the basic phenomena related to the behavior of proteins at various interfaces, the following modification methods are reviewed: attachment of hydrophobic groups, increasing the anionic charges, deamidation and phosphorylation, formation of protein–polysaccharide conjugates, and proteolysis and linking of various functional groups by enzymatic reactions. These chapters are followed by a discussion of noncovalent modifications, such as binding of surfactants to proteins and denaturation of globular proteins. Some of these modifications are presented in the last chapter, which focuses on applications of proteins in food products.

Other methods, such as attachment of positively charged groups to the protein, are not included in the book, mainly because of the lack of information linking modification to specific surface activity.

Some of the modifications discussed here have already been applied in various products. Sometimes they are used unknowingly, for example, in cosmetic products containing both surfactants and proteins.

I hope that the scientific information and the description of the possible approaches to altering functional properties and surface activity of proteins presented in this book will be useful to scientists interested in understanding the behavior of proteins at interfaces, as well as to those who wish to find new applications and products based on proteins.

Shlomo Magdassi

Contents

Contributors

Gyöngyi Hajós   Department of Chemistry II, Central Food Research Institute, Budapest, Hungary

W. James Harper   Department of Food Science and Technology, The Ohio State University, Columbus, Ohio

Malcolm N. Jones   School of Biological Sciences, University of Manchester, Manchester, England

Alexander Kamyshny   Casali Institute of Applied Chemistry, The Hebrew University of Jerusalem, Jerusalem, Israel

Akio Kato   Department of Biochemistry, Yamaguchi University, Yamaguchi, Japan

Jacques Lefebvre   Laboratory of Physical Chemistry of Macromolecules, Institut National de la Recherche Agronomique, Nantes, France

Shlomo Magdassi   Casali Institute of Applied Chemistry, The Hebrew University of Jerusalem, Jerusalem, Israel

M. E. Mangino   Department of Food Science and Technology, The Ohio State University, Columbus, Ohio

Perla Relkin   Department of Food Science, Ecole Nationale Supérieure des Industries Alimentaires, Massy, France

K. D. Schwenke   Research Group Plant Protein Chemistry, University of Potsdam, Bergholz-Rehbrücke, Germany

Frederick F. Shih   Food Processing and Sensory Quality, Southern Regional Research Center, USDA, New Orleans, Louisiana

Ofer Toledano   Casali Institute of Applied Chemistry, The Hebrew University of Jerusalem, Jerusalem, Israel

1
Introduction
Surface Activity and Functional Properties of Proteins

Shlomo Magdassi and Alexander Kamyshny

The Hebrew University of Jerusalem, Jerusalem, Israel

I. INTRODUCTION

In the broad sense, the functional properties of proteins are those affecting their behavior during the preparation, processing, storage, and use of various protein-containing products.

Knowledge of the factors influencing the functional properties of proteins and the ability to regulate these properties are of great importance in developing new processes in the biotechnology, food, pharmaceutical, and cosmetic industries.

The most important functional properties of proteins are solubility, water absorption and binding, rheology modification, emulsifying activity and emulsion stabilization, gel formation, foam formation and stabilization, and fat absorption [1–6]. They reflect the inherent properties of proteins as well as the manner of interaction with other components of the system under investigation.

In other words, the sum of functional properties depends on the physicochemical characteristics of the whole system containing the "working" protein. The determinant properties of the protein itself are the amino acid composition, structure (primary, secondary, tertiary, quaternary), and conformational stability; the charge of the molecule and its dimensions, shape, and topography; the extent of polarity and hydrophobicity, and the nature of protein–protein interactions.

Within certain limits it is possible to change these properties using chemical modification, enzymatic hydrolysis, heat denaturation, etc.

The most important characteristics of the medium influencing the functional properties of proteins are the solvent nature, pH, ionic strength, temperature, redox state, and presence of interfaces. In general, any interface-involving process is likely to be determined by protein binding to this interface, which lowers the surface energy of a heterogeneous system [7–9], i.e., by the surface activity of the protein. The latter is closely related to functional properties and is utilized in many natural processes as well as in many technological applications, mainly in the food, pharmaceutical, and cosmetic industries (e.g., emulsion and foam formation and stabilization). In addition, the biocompatibility of implants, mammalian and bacterial cells, initiation of blood coagulation, solid phase immunoassay, and protein binding to cell surface receptors also involve proteins operating at interfaces.

This chapter describes the origins of surface activity of proteins and some basic characteristics of their adsorption at fluid/fluid and solid/liquid interfaces, which have great importance in various colloidal systems. Understanding the surface activity of proteins would establish the need for various chemical and physicochemical alterations of proteins, which can influence their surface activity. Some of these modifications are presented briefly in this chapter, and in more detail in the following chapters.

II. ORIGINS OF SURFACE ACTIVITY OF PROTEINS

Surface-active materials consist of molecules containing both polar and nonpolar portions, i.e., amphiphilic molecules. The proteins are typically amphiphilic, polymeric substances made of amino acid residues combined in definite sequences by peptide bonds (primary structure). In many cases polypeptide chains are present in helical or β-sheet configuration (secondary structure) which are stabilized by intramolecular (S–S and hydrogen) bonding. The next structural level, the tertiary structure, is determined by the folding of the polypeptide chains to more or less compact globules, maintained by hydrogen bonding, van der Waals forces, disulfide bonds, etc. The globules (subunits) can associate into small clusters (quaternary structure). These features of the protein structure determine surface activity, and the difference in surface activity among proteins arises mainly from variations in their structures.

Any significant alteration in arrangement of the polypeptide chain, without its breaking, is termed denaturation. This process is usually, but not necessarily, irreversible and often takes place when the protein is adsorbed at high-energy air/water and oil/water interfaces. Unfolding of adsorbed molecules allows the polypeptide chains to orient with most of the polar groups in the water phase and most of the nonpolar groups towards the air or oil phase.

The main molecular properties of proteins responsible for their surface activity are size, charge, features of structure, stability, amphipathicity, and lipophylity [2,8]. These parameters will be described briefly.

The size of a molecule is an important feature because proteins form multiple contacts with the surface (e.g., 77 contact points in the case of the albumin molecule and 703 contact points in the case of the fibrinogen molecule adsorbed on silica [10]). "Multipoint" binding usually causes adsorption "irreversibility" having a dynamic nature in the absence of irreversible denaturation. The rates of desorption are, as a rule, much lower than those of adsorption, and in many cases it is virtually impossible to attain the equilibrium state desorbing the adsorbed protein [11]. In other words, the formation of one or several bonds with the surface increases the probability of adsorption of neighboring sites of the same molecule. On the other hand, the desorption of a protein molecule requires the simultaneous rupture of a large number of bonds and, for kinetic reasons, equilibrium is not attained [12–14]. This corresponds to a considerable difference between the activation energies for the adsorption and desorption processes [15,16].

The charge and its density and distribution in the protein molecule strongly influence the surface activity. Experimentally, proteins have frequently been found to exhibit greater surface activity near the isoelectric point, because of minimization of electrostatic repulsion between the identically charged adsorbed molecules [2,8–10,15,17].

This behavior should be especially apparent in the case of an uncharged surface. For adsorption at ionic surfaces the main factor is, probably, the "net" opposite charge of the protein molecule, which may contribute to the enthalpic part of the adsorption free energy. At the same time, a nonuniform distribution of ionic "patches" on the surface of a protein can lead to attractive electrostatic interactions between the patches and the surface even when the "net" charge of the protein is of the same type as that of the surface [18].

The role of *protein structure* in surface activity is still not well understood [8]. The most important factor here is the conformational stability. Depending on the *protein molecule rigidity* and the nature of the interface, proteins can unfold in the interfacial layer or retain their tertiary structure. The proteins with a rigid structure, maintained by intramolecular (e.g., disulfide) bonds, as a rule, retain (at least partially) the native conformation, and structural rearrangements do not contribute to the adsorption process. "Soft" protein molecules undergo structural rearrangements at interface and should be more surface-active than "rigid" proteins, because more contacts with a surface could be formed and because the configurational entropy gain favors the adsorption [8,17,19,20]. Finally, the proteins with *quaternary structure* can display increased surface activity relative to that of structurally analogous monomeric proteins because of the increasing number of contacts with a surface [8].

The chemical differences arising from the differences in the primary structure are also very important because the balance of polar, nonpolar and charged amino acid side chains determines the surface activity of proteins in a particular system, i.e., the possibility and mode of their location at interfaces of different types. This amphipathic nature of the protein molecule allows it to bind with surfaces of different chemical nature. A very important property is the protein *hydrophobicity* [17]. It influences adsorption and orientation of proteins at interfaces and in many cases correlates with surface activity [2,21].

Evidently, in many cases the effect of each factor can hardly be evaluated, and the differences in surface activity of proteins are complex functions of their molecular properties.

III. ADSORPTION OF PROTEINS AT INTERFACES

A. Fluid/Fluid Interfaces

Adsorption kinetics and isotherms. According to the Gibbs equation for ideal systems, the adsorption of classical surface-active molecules at fluid interfaces decreases the surface tension:

$$d\sigma = -\Gamma RT d\ln C_b \tag{1}$$

and

$$\Gamma = \frac{1}{RT}\frac{d\Pi}{d\ln C_b} \tag{2}$$

where σ is the surface tension, Γ is the excess of the surface-active substance (surface concentration) at the interface, C_b is the bulk concentration, and $\Pi = \sigma_0 - \sigma$ is the surface pressure (σ_0 is the surface tension of the solvent).

At low bulk surfactant concentrations, Eq. (1) can be transformed [7] to

$$\Gamma = \frac{\sigma_0 - \sigma}{RT} = \frac{\Pi}{RT} \tag{3}$$

By introducing the values $A = 1/\Gamma$ (molar area of the surfactant in the interfacial layer) and A_m (molecular area), we obtain

$$\Pi A = RT \tag{4}$$

$$\Pi A_m = kT \tag{5}$$

These equations are analogous to the state equations for ideal gases.

The adsorption at fluid interface can be described by at least three consecutive or competitive processes: (1) diffusion of the molecule from the bulk to an interface and attachment to this interface; (2) penetration of new molecules into the adsorbed layer; (3) molecular rearrangement of the adsorbed molecules (this process is very important in the case of proteins). For the last two processes energy barriers exist. If diffusion is the rate limiting process, the rate equations in integrated form according to Ward and Tordai [22,23] are

$$\Gamma = 2C_b \left(\frac{Dt}{\pi}\right)^{1/2} \tag{6}$$

$$\Pi = 2RTC_b \left(\frac{Dt}{\pi}\right)^{1/2} \tag{7}$$

and in differential form

$$\frac{d\Gamma}{dt} = C_b \left(\frac{D}{\pi t}\right)^{1/2} \tag{8}$$

$$\frac{d\Pi}{dt} = RTC_b \left(\frac{D}{\pi t}\right)^{1/2} \tag{9}$$

where D is the diffusion coefficient.

If an energy barrier for adsorption exists, after finite time the adsorption rate can be described by

$$\frac{d\Gamma}{dt} = k_a \exp\left(\frac{-E_a}{RT}\right) C_b (\Gamma_{max} - \Gamma) \tag{10}$$

where k_a is the pre-exponential factor, E_a is the activation energy, and Γ_{max} is the utmost surface concentration. If after formation of a loosely packed monolayer the activation energy is related to the configurational rearrangement of adsorbed molecules, and this rearrangement is stipulated by compression of the interfacial layer against the surface pressure, in order to create an area ΔA_m for the entering molecule, the activation energy is given by $E_a = \Pi \Delta A_m$, and the adsorption rate is given by [15]

$$\frac{d\Gamma}{dt} = k_a \exp\left(\frac{-\Pi \Delta A_m}{kT}\right) C_b \tag{11}$$

In some cases replacement of $d\Gamma/dt$ in equation (11) for $d\Pi/dt$ is useful in determining the ΔA_m values and in interpretation of the adsorption behavior of proteins [22,23]:

$$\ln \frac{d\Pi}{dt} = \ln B - \frac{\Pi \Delta A_m}{kT} \tag{12}$$

where B is a value depending on the bulk protein concentration and on the pre-exponential factor.

In general, the decrease of the adsorption rate after a long period of time is not an evidence of adsorption barrier existence and can simply be a consequence of equilibrium between the surface and sub-surface concentrations, when back diffusion cannot be neglected.

In case of a reversible process, the rate equations for diffusion controlled desorption are given by [24]

$$n_s = 2C_s \left(\frac{Dt}{\pi}\right)^{1/2} \tag{13}$$

$$\frac{dn_s}{dt} = C_s \left(\frac{D}{\pi t}\right)^{1/2} \tag{14}$$

where n_s is the number of molecules desorbing from a unit surface and C_s is the protein concentration in a thin subsurface layer assumed to be in equilibrium with the monolayer.

For energy barrier controlled processes, the general expression for adsorption kinetics is analogous to the Langmuir equation for uniform surfaces:

$$\frac{d\Gamma}{dt} = k_a \exp\left(\frac{-E_a}{RT}\right) C_b (\Gamma_{max} - \Gamma) - k_d \exp\left(\frac{-E_d}{RT}\right) \Gamma \tag{15}$$

where k_d is the pre-exponential factor for the desorption rate and E_d is the activation energy of desorption. Several improved models fitting to experimental adsorption isotherms are described in [25–29].

The available data concerning the adsorption and desorption kinetics and reversibility of the process for fluid interfaces are rather contradictory.

In many cases the adsorption rate is diffusion controlled, at least in the initial stages, with the rate adequately described by Eqs. (6) and (7) at low surface pressures [30–33]. The Γ versus $t^{1/2}$ plots for β-casein and lysozyme adsorption at air/water interface give straight lines at $t \leq 1h$ and remain linear until 10% of the total protein is adsorbed [34]. The diffusion coefficients calculated from the data obtained are $5 \cdot 10^{-10}$ and $8 \cdot 10^{-10}$ m^2/s for β-casein and lysozyme respectively (it should be emphasized that in many cases these diffusion coefficients are higher than the values obtained by ultracentrifuge techniques, probably due to convection effects). At the same time the results obtained in [35] for ovalbumin and lysozyme show that equations (4), (5), and (7) are not applicable, because the more or less linear parts of the $\Pi - t^{1/2}$ plots do not coincide with the linear parts of the $\Gamma - t^{1/2}$ plots. For a number of proteins [15,16,34–37], plotting $\ln(d\Gamma/dt)$ versus Π according to Eq. (11) yields straight lines from which ΔA_m were calculated. These values are much smaller than those expected from the protein molecule dimensions and fall between narrow limits from 50 to 200 Å^2. The Π-A curves yield much higher molecular cross-sectional area [15]. Surface viscosity data also indicate that for protein monolayers the flow unit is much smaller than the molecular size and is within the range from 6 to 8 amino acid residues [38]. Since the surface pressure usually continues to increase even after the surface concentration reaches a constant value, it appears that only conformational changes take place at the last stage of adsorption.

The conclusions that may be drawn so far are (1) that only a small portion of the protein molecule should enter the interface, so that the adsorption could proceed spontaneously, and (2) that the spreading and conformational changes in the protein molecule occur by movement of only small segments of the molecule, several amino acid residues each.

It is likely that experimentally found values of molecular cross-sectional areas do not correspond to the equilibrium states of the protein layers but reflect only the transition state configuration, as assumed by MacRitchie [16] and depicted in Fig. 1. The problem of adsorption reversibility is basic for understanding the protein behavior at interfaces. The belief in the protein adsorption irreversibility is mainly based on drastic conformational changes in interfacial film and the great difficulty of desorbing a protein from this film [15]. However, these criteria are not always a proof of irreversibility. It was shown in many cases [3,24,39–41] that proteins can be desorbed

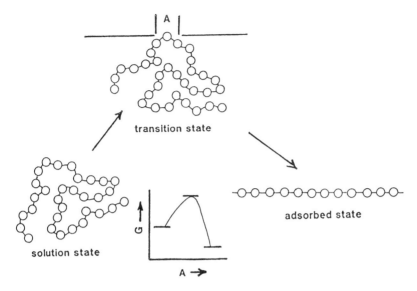

Figure 1 Schematic illustration of the transition state of a protein molecule. The inserted graph shows how the free energy G of the system changes as the area of penetration of the molecule in the surface A increases. The maximum in G corresponds to the critical area for adsorption. (From Ref. 16. Reprinted with permission.)

from monolayers at high surface pressures ($\Pi > 15$ mN/m). Equation (13) describes satisfactorily the experimental data for β-lactoglobulin monolayers at surface pressures of 20, 25 and 30 mN/m [40]. Desorption also occurs on compression of the adsorbed films of acetyllysozyme at not very low bulk concentration; and when after compression the film was expanded to its initial area, both the surface pressure and the surface concentration returned to their values before compression [42].

If the dn_s/dt versus $t^{-1/2}$ plots do not extrapolate to zero at infinite time, as required by a purely diffusion controlled process, this indicates the presence of an activation barrier to the desorption. This behavior was rationalized by MacRitchie [16]. As shown in Fig. 2, most of the segments are located at the interface at low surface pressure. The probability that the transition state configuration could oc-

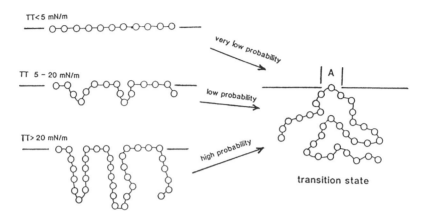

Figure 2 Schematic illustration of how the probability of fluctuation attaining the transition state configuration increases with increasing surface pressure. (From Ref. 16. Reprinted with permission.)

cur by means of fluctuation is negligible. As the surface pressure is increased, the equilibrium between adsorbed and expelled segments shifts towards the latter, and at high surface pressures the probability of a fluctuational attaining of the transition state configuration becomes realistic.

In any case, if the adsorption process is thermodynamically reversible, all points of the adsorption isotherm should correspond to the dynamic equilibrium, and after desorption the protein molecule should revert to its original solution configuration, recovering all of the original properties. At the same time, the experimental isotherms usually fit classical adsorption equations to a limited extent.

It was shown in the detailed investigation of Graham and Phillips [34,36,43] that the Γ-C_b isotherm for β-casein at air/water interface revealed a well-defined plateau over the wide concentration range. However, at $C_b > 10^{-2}$ wt % the film thickness and Γ increase further (without any significant changes in Π) due to the presence of reversibly adsorbed molecules. This reversibility can be connected with either formation of second and subsequent layers [36] or molecular reorientation of adsorbed β-casein molecules [26]. The adsorption isotherms for lysozyme and bovine serum albumin (BSA) do not exhibit a plateau in surface concentration in the region where the surface pressure increases. All proteins reduce the interfacial pressure more effectively when the apolar phase is oil. The shapes of isotherms for β-casein and BSA at toluene/water interface are similar to those at air/water interface. At the same time the isotherms for lysozyme are different, with an inflection point in the Π (air/water) $-$ C_b plot at 8 mN/m that is absent in the Π (toluene/water) $-$ C_b plot [36]. The two-step adsorption isotherm was also obtained for RNase [44]. The enzyme activity can be recovered from the films of pepsin and trypsin, and this was regarded as evidence for the presence of native or incompletely unfolded molecules [12]. In general, it is likely that a surface saturated with protein at high bulk concentrations is composed of a mixture of native and denatured molecules, which are irreversibly adsorbed. From the available data concerning protein adsorption at fluid/fluid interfaces, the generalized conclusion can be drawn that in many cases the experimental adsorption isotherms are only apparent isotherms, not corresponding to true thermodynamic equilibrium.

The structure and molecular properties are the basic factors influencing the interfacial behavior of proteins. The role of the *molecular shape* and *rigidity* becomes apparent when comparing the surface activity of three well-defined proteins: β-casein (random coil with no disulfide bonds and high content of nonpolar amino acid residues), lysozyme (compact globule with tertiary structure maintained by disulfide bonds), and BSA (also a globular protein containing 17 intramolecular S–S bridges; its unfolding proceeds easier than the lysozyme unfolding). At the same concentrations the disordered β-casein is more effective in surface tension reduction than the globular BSA and lysozyme [34,36,43], and it reaches the steady state Π and Γ values faster. The Π-A_m curve for β-casein can be described in terms of segments of a disordered polypeptide chain forming either trains of amino acid residues at the interface or loops and trains of residues protruding into the bulk phase. In contrast, the globular lysozyme molecules in condensed films retain a structure similar to the native one. The structure of BSA at the interface is intermediate between those of lysozyme and β-casein (Fig. 3). The maximum amounts of irreversibly adsorbed β-casein, lysozyme, and BSA at air/water inter-

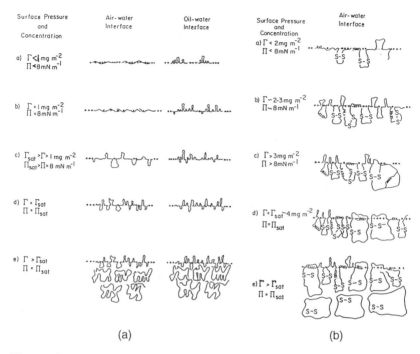

(a)   (b)

Figure 3 Schematic representation of the structure of adsorbed films of β-casein (a) and lysozyme (b) at fluid/fluid interfaces. (From Ref. 43. Reprinted with permission.)

face are 2.6, 3.0, and 3.3 mg/m^2 respectively [36]. The interfacial films formed by globular lysozyme and BSA have more rigid structure with high shear viscosity and elasticity [3]. The influence of shape and flexibility of the protein molecules becomes distinctly apparent at competitive adsorption. The flexible, rod-like molecules of β-casein, when coadsorbed with lysozyme, can exchange with β-casein in solution, while rigid globular molecules of lysozyme cannot. These physical features of the two proteins also seem to allow β-casein to adsorb in intermolecular spaces, which lysozyme cannot access when both proteins are present [45].

Unlike for β-casein, the Π-A_m curves for BSA and lysozyme are more expanded at the oil/water than at the air/water interface [43].

This suggests that both proteins adopt more extensively unfolded conformation at the oil/water interface in consequence of apolar side chain solvation by oil molecules. These data testify also to the importance of effective *hydrophobicity* of proteins for their surface activity. In general, a good correlation exists between these properties. For instance, the hydrophobicity indices for β-casein, BSA, and lysozyme according to Bigelow [46] are 1320, 1120, and 970 cal/mol residue respectively. Keshavarz and Nakai [47] and Lee and Kim [48], while studying the interfacial tension of several proteins at water/corn oil and water/soybean oil interfaces, showed that the more hydrophobic the protein, the greater the depression in the interfacial tension. Protein unfolding by heat or by urea treatment leads to increase in its surface hydrophobicity and hence surface activity. After lysozyme denaturation at 50°C and pH 1, its adsorption isotherm and the Π-C_b plot have forms reminiscent of those for disordered β-casein. This increase in surface activity is a consequence of the exposure of hydrophobic amino acid residues that would normally be located in the interior of the globular molecule [43]. The increase of surface hydrophobicity of ovalbumin [21] and BSA [49], after their denaturation, also leads to greater reduction of interfacial tension. Changes in hydrophobicity and hence surface activity of proteins can also be obtained by chemical modification of proteins. It was demonstrated by Magdassi et al. [9,50] that after covalent attachment of hydrophobic fatty acid chains to ovalbumin and IgG the surface tension was reduced faster and to a lower value compared to the native proteins (Fig. 4). Glycosylation of β-lactoglobulin with maltose leads to more open tertiary structure and some disordering of the protein molecule [2,51]. As more maltose groups were bound, the surface hydrophobicity was decreased and net negative charge was increased, which causes a reduction of adsorption rates. In addition, the carbohydrate groups restrict close molecular packing in the interfacial film and inhibit unfolding of the hydrophobic regions to the apolar phase. When the β-casein molecule was chemically modified by acyl groups (acetyl, *n*-butyryl, *n*-hexanoyl, *n*-decanoyl), an increase of the molar surface area at high surface pressures was observed (Fig. 5 [52]), and this effect may be connected not only with the alteration of the protein hydrophobicity but also with the electrostatic repulsion of nega-

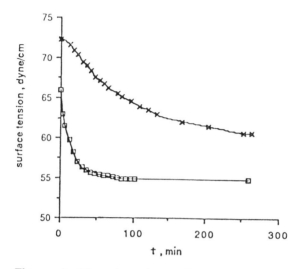

Figure 4   Time dependence of surface tension for native ovalbumin (x) and ovalbumin with covalently attached caproyl groups (□) at pH 10. (From Ref. 9. Reprinted with permission.)

tively charged molecules (at pH 7.0 the charge of the β-casein molecule is -11 and that of acyl derivative -19).

Generally, the *surface charge* of protein molecules, together with hydrophobicity, strongly influences their adsorption at fluid/fluid interfaces. The observations that the highest adsorption for a number of proteins takes place near their isoelectric points [3,9,30,53,54] result from the decreased electric barrier for adsorption, decreased repulsion in interfacial layer, and minimum solubility of the proteins in the bulk phase. On the contrary, according to data of Graham and Phillips [43], the adsorption of β-casein, BSA, and lysozyme is pH-independent in a wide range of pH including isoelectric points (5, 4.5, and 11 respectively), and for β-casein and BSA the surface concentration decreases only at pH values sufficiently high to induce a large negative charge on the protein molecules. It indicates that the hydrophobicity contribution to the adsorption-free energy is dominant.

There are not many reports about the influence of *size* and *molecular weight* on the surface activity of proteins. It was shown [55] that sodium caseinate and proteins from whey concentrate diffuse to the

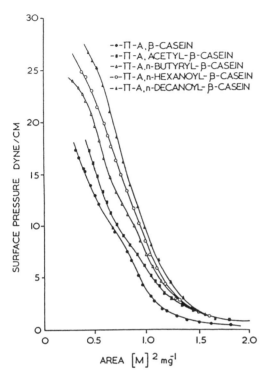

Figure 5 Π-A isotherms of β-casein and its acyl derivatives on phosphate buffer (pH 7) subphase at 25°C. (From Ref. 52. Reprinted with permission.)

air/water interface faster than soy isolate proteins, having a higher molecular weight and complex quaternary structure. In the mixed monolayers of insulin with γ-globulin or catalase, insulin can be completely desorbed at suitable surface pressure. The absolute rate of desorption is the same as in the case of pure monolayers, providing the area occupied is over 30% of the total surface area [40].

It is very likely that any attempt to correlate the molecule size with the surface activity from available data on native proteins will fail due to the dominant role of the structural parameters. We expect that such correlations would be obtained only by using homologous series of rodlike proteins, which can be obtained at various molecular weights. Such an approach is being evaluated now in the authors' laboratory for enzymatically modified gelatin.

In summary, it appears that the adsorption of proteins at fluid/fluid interfaces is very complex and involves several steps that are not encountered in classical surfactants or even polymeric surfactants. All the parameters that are controlling the surface activity may be changed by chemical or physicochemical modifications, such as covalent attachment of various groups, breaking of disulfide bonds, denaturation by heat, etc. It is important to note that while changing one parameter, one may change the others as well, and therefore it would be difficult to isolate definite parameters.

B. Solid/Liquid Interfaces

A large number of applications in medicine, biology, and biotechnology are based on the adsorption of proteins onto solid surfaces—the simplest method of their immobilization. Many chromatographic separations, such as hydrophobic, displacement, and ion-exchange chromatographies, are based on different affinities of proteins for the support. Protein adsorption at implanted biomaterials is believed to play an important role in determining their biocompatibility with various biological systems. Another demonstration of the importance of protein adsorption onto solid surfaces is the extensive use of immunodiagnostic reagents, in which antibodies are adsorbed on latex or gold particles or on polystyrene plates.

Adsorption at solid/liquid interfaces has some peculiarities as compared with fluid/fluid interfaces. *The chemical nature* of the solid surface and its properties (charge, hydrophobicity, etc.) determine the mode and strength of binding, as well as, in many cases, the conformational changes in adsorbed protein molecules. The solid surfaces can be easily modified and "tuned up" for specific types of interactions. Usually, in contrast to fluid surfaces, solid surfaces are not chemically or energetically uniform, and their heterogeneity may result in nonuniform adsorption of protein layers. Finally, adsorption from solutions is always a competitive process, and in the simplest case competition between a protein and a solvent takes place.

A large number of different carriers for protein adsorption have been described: glass [12,13,56–61], silicas [10–13,61–66], clays [12,13,56,57,67], insoluble inorganic salts [13,56,68–71], ceramics

[72], metals and metal oxides [13,17,56,58,61,66,70,73–76], molecular sieves [6], carbon [11,66,77,78], many types of neutral and charged polymers [12,13,17,59,61,63,70,71,73,79–84], and carriers of mixed types, obtained by deposition of lipid layers (e.g., lecithin and cholesterol) on a solid support [11].

Adsorption kinetics and isotherms. The rate of protein adsorption onto solids is usually much slower than that predicted from the diffusion theory [85–87]. For various protein-adsorbent systems, the period of time required to obtain maximum adsorption ranges, as a rule, from several tens of minutes [10,12,14,88] to several hours [11,12,14,63,65,66,79,81,84,89,90]. More rarely, the adsorption terminates after several minutes [67,91] or continues for 24 h and longer [92,93]. It cannot be excluded, however, that the initial adsorption rates should be transport limited, as has been shown by Norde et al. [94] for adsorption of lysozyme, RNase, and myoglobin on glass. The importance of diffusion is also obvious at the first step of adsorption from protein mixtures [95]. In this case the interface accommodates initially the protein molecules with the largest diffusion coefficients, and afterwards these molecules may be displaced by other molecules with higher affinity to the surface.

As in the case of fluid interfaces, the question of whether the adsorption of proteins onto solids is reversible or irreversible is very important for correct estimation of physicochemical characteristics of the process. In a reversible process, dilution of sorbate in the bulk phase should lead to spontaneous desorption of some portion of adsorbed molecules up to elimination of a transient difference in the chemical potential of the sorbate at the interface and in the solution; the "ascending" and "descending" branches of the isotherm must overlap at all values of C_b. Only in this case the isotherm represents thermodynamic equilibrium, and the equilibrium constant K_{ads} and the standard Gibbs energy of adsorption $\Delta G^0_{ads} = \Delta H^0_{ads} - T\Delta S^0_{ads}$ can be determined.

The adsorption of proteins from aqueous solutions is often described in terms of the Langmuir equation

$$\Gamma = \frac{K_{ads}\,\Gamma_{max}\,C_b}{1 + K_{ads}C_b} \tag{16}$$

with fairly rapid attainment of a well-defined plateau and high affinity to a solid [9–13,17,59,66,68,79–81,84,90–92,96]. Sometimes the adsorption data formally obey the Freundlich equation

$$\Gamma = aC_b^{1/n} \tag{17}$$

where the constant a is a measure of adsorption capacity and $1/n$ is a measure of affinity [59,83,97,98]. In some cases the protein adsorption isotherms are complex with maxima [92], S-shaped sections, and inflection points [17,100–102].

Data on the reversibility of the protein adsorption process and thus the applicability of different models for its description are contradictory. Norde et al. [17] consider protein adsorption as a nonequilibrium irreversible process, whose description should be, generally, based on the laws of irreversible thermodynamics.

This opinion is founded on the observations that in many cases either adsorbed proteins cannot be desorbed from surfaces [12,13,17,19,20,64,70,103,104] or hysteresis between the ascending and descending branches of isotherms exists [19,20,61]. In comparison, reversible isotherms were obtained for adsorption of cytochrome c [11], lactate dehydrogenase [105], and catalase [106] on silica, BSA on QAE-dextran [83], glucose-6-phosphate dehydrogenase, alkaline phosphatase, and lactate dehydrogenase on polyaminostyrene [107], and pancreatic lipase on siliconized glass [108]. In addition, exchange of the protein between surface and solution was detected by using labeled proteins [60,87,109–112]. In some cases the apparent irreversibility may be connected with a tendency of the adsorbed proteins to denature over a long period of time. Thus denaturation of cytochrome c adsorbed on various surfaces becomes appreciable after 48 h [11]; the ability to remove fibrinogen and albumin from a variety of polymers [59,113] and albumin from hydrated metal oxides [76] becomes more difficult if the time between adsorption and elution was lengthened. Frequently, the adsorbed protein can be readily desorbed from solids by changing the pH or ionic strength [11,60,70,103] or by using detergents [70,103,114,115]. Finally, if adsorption were truly irreversible, one would expect that the "isotherm" would rise instantly to the plateau value or that in very diluted systems complete depletion would be observed. This does not occur.

Taking these data into account, we may assume that the adsorption of proteins (especially of those with the rigid globular structure) is usually inherently reversible, and the irreversibility is only apparent and has kinetic origins (see above). This provides the possibility to estimate from the experimental isotherms some physicochemical characteristics of the protein adsorption process, such as adsorption equilibrium constants K_{ads} and corresponding standard free energies ΔG^0_{ads} (or their lower limits, if the irreversible entropy and enthalpy changes cannot be completely ignored).

Initial high slopes of adsorption isotherms indicate, usually, a high affinity of proteins for the solid/water interfaces (Fig. 6). The ΔG^0_{ads} values calculated from the Langmuir isotherms are usually in the range between -6 and -12 kcal/mol for various protein-adsorbent

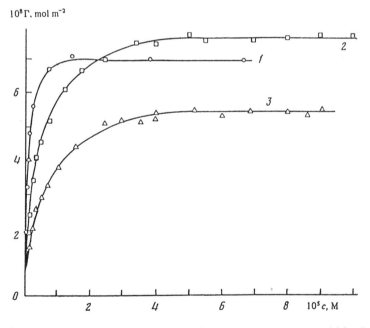

Figure 6 Adsorption isotherms of cytochrome c on graphitized thermal black (1) and cholesterol (2) and lecithin (3) monolayers, deposited on silica gel, at pH 7 and ionic strength 0.2. (After Ref. 11. Reprinted with permission.)

systems [11,90,92,105,106,108,116,117]. It is interesting to note, for example, that the K_{ads} and ΔG^0_{ads} values for adsorption of cytochrome c on silochrome calculated from the linear form of the Langmuir equation are in very good agreement with those calculated from kinetic data [11].

A detailed analysis of different contributions to the enthalpy and entropy changes, for a number of protein-adsorbent systems, was carried out by Norde, Lyklema, and others [17,103,118,119]. They showed that the protein adsorption on solids is often endothermic and that the driving force of the process is the positive ΔS_{ads}. The main contributions to large positive values of entropy can arise from sorbent surface dehydration and from the conformational rearrangement of protein molecules upon adsorption [17].

In general, the plateau values Γ_{max} of experimental isotherms give ambiguous information about the conformational state of the adsorbed protein molecules. In theory, maximum surface concentrations for unperturbed globular proteins range from about 1.5 to 8 mg/m^2 [17]. In some cases the plateau values of the isotherms correspond roughly to a close-packed monolayer in accordance with known dimensions of the protein molecules [80]. However, in many cases it was observed that the surface concentrations at saturation are considerably lower than the expected values based on molecular size [4,10,103,108,116,117]. For instance, the cytochome c molecule has an ellipsoidal shape with dimensions $30 \times 34 \times 34$ Å. This yields a circular cross section area of ~ 1000 Å^2 for hexagonal close packing and ~ 1150 Å^2 for cubic packing and corresponds to $\Gamma_{max} \sim 2$ mg/m^2. The experimental values of the molecular areas are significantly larger, ranging from 4150 Å^2 ($\Gamma_{max} \sim 0.5$ mg/m^2) for silochrome at pH 5 to 2000 Å^2 (Γ_{max} 1 mg/m^2) for silica gel at pH 7 [11]. This phenomenon can be explained by either (1) formation of a nondense ("diluted") monolayer or (2) unfolding of a protein molecule, in a similar way to that described for fluid/fluid interfaces. Although the characteristics of a solid/liquid interface (hydrophobicity, polarity, sign of charge and its distribution, degree of uniformity) are very important, it also appears that the *structure and molecular properties* of the protein play the dominant role in the state of the interfacial protein layer [9,103]. There is now substantial evidence that proteins undergo structural alterations at the solid/water interface. Kondo et

al. [62] have found that the magnitude of structural changes in protein molecules (cytochrome c, RNase A, myoglobin, ovalbumin, he-
moglobin, BSA) upon adsorption on ultrafine silica particles, as estimated by measuring the circular dichroism spectra, increased with
the increase in their adiabatic compressibility, i.e., decreases in the
molecular rigidity, and also depends on the affinities of the particles
for proteins. α-Lactalbumin, which has a relatively low native-state
stability, retains virtually none of its secondary structure upon adsorption on α-Fe$_2$O$_3$. On the contrary, stable globular lysozyme retains most of its native structure upon adsorption on this adsorbent
[17]. The loss of α-helix content for lysozyme adsorbed on silica at
monolayer coverage is $\sim$3–7%, whereas flexible human serum albumin loses $\sim$36% of the secondary structure [120]. For fibrinogen
eluted from glass, the loss of α-helix content is about 50% [121].
The absorption spectrum of very rigid small globular ferricytochrome
c desorbed from silica coincides completely with that of the native
protein, including the band at 695 nm, which is very sensitive to
conformational changes. Haynes and Norde [17] have shown that for
proteins adsorbed on negatively charged polystyrene, Γ_{max} was large
for the least stable α-lactalbumin and decreased with the increase of
structure stability (RNase > myoglobin > superoxide dismutase). Lysozyme, however, diverges from this trend, possibly because of its
tendency to aggregate in solution and on surfaces.

In some cases the degree of protein unfolding upon adsorption depends on the surface concentration. Morrissey and Stromberg [10]
showed that the fraction of γ-globulin carbonyl groups bonded to
colloidal silica was higher (ca. 0.20) when adsorption occurred from
diluted protein solution than from a concentrated one (ca. 0.02 in the
plateau region of the isotherm), i.e., the degree of unfolding decreased with the increase in surface coverage. The dependence of the
conformation of globulins on the degree of the surface coverage has
also been suggested in other studies [122,123]. Although the decrease
of α-helix content with the decrease in surface concentration was observed even for "rigid" lysozyme adsorbed on silica [120], such dependence is not common for different protein–carrier systems.

The surface properties of the protein molecules, especially *hydrophobicity,* as well as the degree of the sorbent surface hydrophobicity, strongly affect the protein adsorption behavior. Calculated

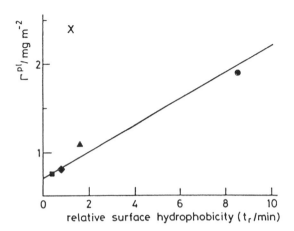

Figure 7 Correlation of plateau values with protein surface hydrophobicity (from hydrophobic interaction chromatography data) for the adsorption of egg-white lysozyme (●), bovine pancrease ribonuclease (▲), α-lact-
albumin (x), sperm whale myoglobin (♦), and superoxide dismutase (■) on
negatively charged polystyrene in 50 mM KCl at 25°C and pH equal to pI
of each protein. (From Ref. 17. Reprinted with permission.)

Gibbs energies of adhesion were found to increase with the increase
in protein surface hydrophobicity, regardless of the relative hydrophobicity of the sorbent [17]. Data shown in Fig. 7 support this conclusion (divergence for α-lactalbumin from this correlation may be
connected with unusually low structural stability of this protein compared with the other proteins studied). Increasing the hydrophobicity
of ovalbumin by acylation of lysine ϵ-amino groups with stearyl residues leads to significant increase in its adsorption on hydrophobized
silica as compared to the native protein and reaches a constant value
at the modification degree above 33% [90]. In many cases, larger
adsorption occurring at hydrophobic interfaces than at hydrophilic
ones is a result of a higher free energy of the process [17,61,80,124].
Human serum albumin adsorbed progressively more onto surfaces
with higher hydrophobicity (adsorption was studied as a function of
surface density of octadecyldimethylsilyl chains on silica). When the
fractional surface coverage of hydrophobic chains was larger than

0.42, the adsorption saturation of protein leveled off [125]. Elwing et al. [126] have prepared surfaces with a gradient of alkyl groups, and using ellipsometry have noted a difference in the amount of fibrinogen and γ-globulin adsorbed at the hydrophobic end of the gradient (0.7 and 0.55 μ/cm^2, respectively) and at the hydrophilic end (0.3 μ/cm^2 for both proteins). In some cases, correlations between protein adsorption and sorbent hydrophobicity indicate that proteins show maximum affinity for surfaces of intermediate polarity [17,127]. For example, under the same conditions, cytochrome c has the highest affinity for partly oxidized graphitized thermal black compared to silica and hydrophobic lecithin and cholesterol monolayers deposited onto this carrier [11].

When a protein is adsorbed on a solid surface with high hydrophobicity, considerable conformational changes, due to hydrophobic interactions, can take place. Andrade et al. [128] observed that the conformation of fibronectin does not change upon adsorption onto hydrophilic silica but changes significantly upon adsorption onto chemically modified hydrophobic silica. Steadman et al. [64] have found that decreasing the surface polarity (methyl-, butyl-, diphenyl-, and carboxy-sulfone-silica) produced decreasing thermal stability and increasing structural alterations of the adsorbed lysozyme. The activity losses by enzymes adsorbed onto hydrophobic polymer surfaces are, in general, larger than those by enzymes adsorbed onto hydrophilic ones [98].

Protein adsorption at the solid/liquid interface is influenced by the *surface charge of the molecule* and the *surface charge of the solid.* The overlap of the electrical double layers at the sorbent and the protein surfaces results in electrostatic attraction if they have opposite charges and in repulsion if their charges have the same sign. Besides, of great importance is the repulsion between the protein molecules in adsorbed layers, which is minimized at their isoelectric points.

On positively charged polystyrene at pH 7 the Γ_{max} values increase in the order lysozyme $<$ RNase $<$ myoglobin $<\alpha$-lactalbumin (α-lactalbumin is the only protein having a net negative charge at these conditions) in accordance with their net electrostatic attractions (or repulsions) to the surface [17]. On negatively charged polystyrene the relative positions of the plateau values are nearly reversed, but

α-lactalbumin also has a relatively large Γ_{max} despite being repelled from the surface. Moreover, the plateau value for RNase is similar on the two surfaces [17]. These observations show that although electrostatic forces are very important, they do not dominate the adsorption process. In many cases Γ_{max} is at a maximum value near the isoelectric points of the protein/sorbent complex [9,10,14,58,70,83,103,129–132]. One possible explanation of this behavior is that increased lateral electrostatic repulsion of the equally charged protein molecules prevents the formation of close-packed monolayers. Another explanation takes into account that the structural stability of the protein molecules is maximal and the conformational rearrangements are minimal when adsorption proceeds at a pH equal to the pI of the protein/sorbent complex [9,17]. Brooksbank et al. [82] gave an example of "tuning" the electrostatic interactions and the structure of the adsorbed protein layer. They showed that most of the β-casein molecule, adsorbed onto negatively charged polystyrene latex particles, lies close to the surface, leaving the highly negatively charged terminal portion to form a loop extending into the solution. Moderating the electrostatic repulsion by diminishing the protein charge via dephosphorylation, or by calcium binding, or by increasing the ionic strength, causes the loops to relax and the β-casein molecules to pack more tightly.

As to the size and the molecular weight of the protein molecules, it seems that they are not important factors in the protein adsorption at solids, and no systematic study of these parameters is available. It may be generally predicted that if all other conditions are kept constant, there will be a tendency for larger protein molecules to be preferentially adsorbed [8].

In many practical applications, e.g., in the food industry or in the creation of biocompatible implants, it is very important to know the adsorption behavior of *multicomponent protein mixtures*. Adsorption from mixtures usually proceeds as a competitive process, and all the above-mentioned factors influence relative preference in adsorption of various proteins. In general, it was found that certain proteins may be enriched at the interface, relative to the others [8,59–61]. Mixtures containing two or three proteins have been used as simple models of complex biological fluids, like plasma [59,60,133] or milk

[134]. However, Brash et al. [60,135,136] have shown that data based on single protein or simple mixture studies cannot always be used to predict protein adsorption from complex mixtures. For example, fibrinogen is preferentially adsorbed on glass from a mixture with albumin and IgG, but it is not detected on any of the hydrophilic surfaces in contact with undiluted plasma. At the same time, dilution of the plasma leads to a gradual increase in fibrinogen adsorption, the maximum being located at the human plasma concentration of about 1% of normal. It was concluded [60] that beyond the maximum the decrease of adsorption is due to displacement of fibrinogen by other plasma components. It is also important to emphasize that by changing solution conditions, e.g., ionic strength or pH, one may affect the properties of only some of the proteins in the mixture and hence decrease or increase the adsorption tendencies depending on the overall physicochemical effects.

IV. EMULSIONS AND FOAMS: FORMATION AND STABILIZATION

Emulsions are dispersed immiscible droplets (oil or fat) within another liquid, which are stabilized by a layer of surface-active molecules. Foams are coarse dispersions of gas in liquid. At foam formation water molecules surround the gas bubbles and tend to arrange them in order, resulting in a high surface energy. As the foam ages, water drains and the air cells approach each other. However, the presence of an interfacial layer, as in the case of an emulsion, provides increased stability even after considerable drainage has occurred. Although, from a physical point of view, there are several quantitative differences between emulsions and foams [137], good emulsifying agents are, in general, also good foaming agents, since the factors influencing emulsion and foam formation and their stability are somewhat similar. Proteins are widely used in commercial food emulsions (e.g., mayonnaise, butter) and foams (e.g., meringues, whipped cream). The factors governing the formation and stabilization of the protein-based emulsions and foams are principally related to those determining the surface activity of proteins at fluid/fluid interfaces and discussed above [2–4,15,24,51,54,55,138–142]:

1. Intrinsic protein properties: size, shape, flexibility, surface charge, hydrophobicity, solubility
2. Adsorbed protein film properties: thickness, rheology (viscosity, cohesiveness, elasticity), net charge and its distribution, degree of hydration
3. Solution conditions (pH, ionic strength, temperature)
4. Processing parameters (shear forces, temperature, phase composition and viscosity, droplet size)

The influence of chemical and physicochemical modification of poteins on their emulsifying and foaming activity has been comprehensively reviewed [1–5,25,50,51,143–148], and we present here only a few examples. Kato et al. [149] showed that an increase in the *rigidity* of lysozyme and BSA molecules reduced their foaming and emulsifying activities as well as foam and emulsion stabilities. At the same time, reduction of the disulfide bonds of BSA results in a more expanded conformation and leads to a decrease in emulsifying activity similar that of the native protein [150]. A complete disruption of the tertiary and secondary structure of BSA in 8 M urea destroyed its emulsifying activity [150]. Schmidt et al. [151] also observed that reducing agents decrease foamability of whey protein concentrate. On the other hand, foam stability was significantly increased with the extent of disulfide bond reduction by dithiothreitol in the glycinin molecule [152]. These results indicate the existence of an optimal structure of a protein molecule for formation of a strong, cohesive interfacial film.

Heat denaturation, leading to increased *surface hydrophobicity* of proteins, usually enhances emulsifying and foaming properties. Therefore partial denaturation of whey proteins at 40 to 65°C improves their functionality in the foam formation process [153]. Preparation with increased foaming and emulsifying activity was obtained after heating the yeast protein at pH 11,8 [154]. Magdassi et al. [155,156] showed the enhancement of emulsifying activity of ovalbumin after its hydrophobization by covalent attachment of alkyl groups to lysine residues with the use of *N*-hydroxysuccinimide esters. The emulsions obtained were also more stable in the presence of modified ovalbumin compared to the emulsions prepared by native protein.

Acetylation and succinylation are the widely studied derivatizations of ϵ-amino groups of lysine residues, increasing the net negative *charge* of the protein molecules and thereby increasing the electrostatic repulsive forces, both intra- and intermolecular, thus favoring greater emulsion and foam stability. Furthermore, acylation causes molecular expansion of the protein molecules, favoring improved functionality [147]. The increase in emulsifying activity was shown for acetylated globin [157], succinylated BSA [150], and yeast and soy proteins [158,159]. The foaming activity of yeast and soy proteins was also improved after this modification [158,159]. Other ways to alter the net charge and amphipathic nature of protein molecules are phosphorylation, dephosphorylation, and deamidation [144,147]. Emulsions, prepared with β-lactoglobulin, phosphorylated by phosphorous oxychloride, were more stable at pH 7 and 5 than those stabilized by the native protein [160]. Phosphorylation of yeast proteins leads to enhanced emulsifying and foaming properties [161]. At the same time, emulsifying properties of phosvitin were decreased by removal of the phosphate residues by phosphatase [5]. Deamidation of proteins usually results in improved functionality [5].

The main consequence of glycosylation by neutral sugars is the increased hydrogen bonding for both protein–protein and protein–solvent. The more flexible glycosylated derivatives, by virtue of their greater hydration and enhanced ability to act as steric stabilizers, would be expected to display improved emulsifying and foaming properties [147]. The corroboration of this idea was obtained in the case of maltosyl-β-lactoglobulin [162].

One of the current approaches to the improvement of the functional properties of proteins is enzymatic hydrolysis [148]. The emulsifying ability of soy protein isolate can be increased by treatment with neutral fungal protease; however, this treatment decreases emulsion stability [163]. Partial hydrolysis of fish protein concentrate improves both emulsification and stability [164]. On the other hand, treatment of whey protein concentrate with pepsin, pronase, and prolase leads to a decrease in emulsification ability, suggesting that there is an optimal molecular size of proteins contributing to emulsification [165]; the specific foam volume was increased by very limited hydrolysis but decreased by a more extensive process, while foam stability was greatly decreased even by limited hydrolysis [165,166]. Foam volume and stability of soy proteins, partially digested by rennin, were superior in comparison with native proteins [167]. Partial hydrolysis of peanut proteins with papain significantly increased both foaming capacity and foam volume [168]. Pepsin and papain hydrolysates of rapeseed protein concentrate displayed increased foam volumes and decreased drainage, compared to untreated control [169].

Treatment of albumin with trypsin, bromelain, and fungal protease produced significantly larger volumes of foam as compared to control. However, treated protein displayed lower foam stability [170].

Recent achievements in emulsion and foam preparation and application are described in detail in the chapter written by Mangino and Harper.

V. GELATION IN PROTEIN SYSTEMS

Generally speaking, gelatin is not a surface phenomenon, but it is an important functional property of proteins. Protein gel represents a three-dimensional network of considerable rigidity and elasticity in which water is either tightly bound to polar amino acid residues or more loosely associated within the interstitial cells [4]. According to Ferry [171], protein gel formation is considered as a two-stage process involving initial denaturation of macromolecules followed by aggregation:

$$xP_n \rightarrow xP_d \longleftrightarrow (P_d)_x$$

where n denotes the native state and d the denaturated one. The denaturated molecules can orient themselves to a certain degree of order before aggregation. Barbu and Joly [172] and Kratchovil et al. [173,174] proposed that partially unfolded globular protein molecules aggregate in linear associates when repulsion is large, and in random ones when repulsion is small, e.g., at the isoelectric point. The statistical approaches to the protein gelation process are hardly applicable, because the protein molecule surface has a mosaic-like structure with sites differing in charge density and sign, degree of hydrophobicity, and presence of disulfide groups [175]. The present view [175,176] supports the idea that partially unfolded globular proteins associate linearly at low temperatures and randomly at high temperatures and at the isoelectric point.

For gel formation and its stability, a balance between the attractive forces and the repulsive forces is necessary [3,177,178]. This balance is determined by the molecular properties of protein (structure, size of molecules, hydrophobicity, surface charge and distribution) as well as processing parameters (temperature, heating and cooling rate, pH, ionic strength, presence of chelating ions, etc.). The main forces responsible for gel formation are hydrogen bonding, electrostatic and hydrophobic interactions, and disulfide cross-links [3]. Cross-linking is very important for gel formation and stabilization, and together with the fluidity of the solvent inside the gel network it imparts to the gel its elasticity and strength [3,179,180]. Hydrophobic interactions are also very important, especially at the initial stages of the gelation process [3,181–185].

Electrostatic forces can provide attraction between oppositely charged amino acid residues of adjacent parts of a polypeptide chain as well as cause mutual repulsion between residues with the same charge. The importance of the electrostatic interactions is reflected in the effects of pH and salts on gel properties [3,186–189]. In many cases, the formation of gels with necessary characteristics can be achieved only when optimizing all these types of interactions, and the main tools for this optimization are chemical and physicochemical modifications of the protein molecules.

VI. PROSPECTS

There are still many unsolved problems that are extremely important for understanding the mechanisms governing the surface activity of proteins. Numerous investigations deal only with the quantity of adsorbed proteins under various conditions, while many cases lack data on physicochemical characteristics of the surface layer and the functional effects of adsorption. Many modern physical methods do not provide for complete and unequivocal information on the qualitative state of proteins at the interface, and theoretical models describing various aspects of adsorption are approximate and idealized.

The most significant and promising trends in research on the surface activity of proteins are:

Further investigation into the interconnections between molecular properties of proteins and their surface activity.

Determination of general regularities that govern the functional activity of proteins in surface layers.

Thorough study of states and properties of protein layers at various interfaces and development of new methods for analyzing the interfacial state.

Elucidation of behavior of multicomponent protein systems and development of models for competitive adsorption, based on the knowledge of molecular properties of the components, surface characteristics, and medium features.

Combining specific biological activity of various proteins with their surface activity at various interfaces.

Development of methods for control of surface activity by means of chemical and physicochemical modifications of the protein molecules. These modifications may help to elucidate the basis of protein functionality and should enable the design of better protein derivatives with the needed properties.

All these theoretically interesting issues are also very important for the elaboration of new processes in biotechnology, immunology, and medicine, as well as in the food, pharmaceutical, and cosmetic industries.

The following chapters will deal with the various possibilities of changing the surface activity of proteins and may also provide simple tools for systematic studies on the effect of various isolated parameters on surface activity.

REFERENCES

1. A. Pour-El, in *Protein Functionality in Foods* (J. P. Cherry, ed.), ACS Symposium Series, Vol. 147, Washington, D.C., 1979, pp. 1–19.
2. J. E. Kinsella, in *Food Protein Deterioration, Mechanisms and Func-*

tionality (J. P. Cherry, ed.), ACS Symposium Series, Vol. 206, Washington, D.C., 1982, pp. 301–325.

3. P. A. Morrissey, D. M. Mulvihill, and D. O'Riordan, in *Developments in Food Proteins*, Vol. 7 (B. J. F. Hudson, ed.), Elsevier Applied Science, London, 1991, pp. 125–166.

4. K. L. Flinger and M. E. Mangino, in *Interactions of Food Proteins* (N. Parris and R. Barford, eds.), ACS Symposium Series, Vol. 454, Washington, D.C., 1991, pp. 1–12.

5. A. Kato, ibid., pp. 13–24.

6. S. Nakai, E. Li-Chan, M. Hirotsuka, M. C. Vazquez, and G. Arteaga, ibid., pp. 42–58.

7. D. J. Shaw, *Introduction to Colloid and Surface Chemistry*, Butterworths, London, 1985, pp. 71–107, 232–244.

8. T. A. Horbett and J. L. Brash, in *Proteins at Interfaces. Physicochemical and Biochemical Studies* (J. L. Brash and T. A. Horbett, eds.), ACS Symposium Series, Vol. 343, Washington, D.C., 1987, pp. 1–33.

9. S. Magdassi and N. Garti, in *Interfacial Phenomena in Biological Systems* (M. Bender, ed.), Surfactant Science Series, Vol. 39, Marcel Dekker, New York, 1991, pp. 289–299.

10. B. W. Morrissey and R. R. Stromberg, *J. Colloid Interface Sci. 46*: 152 (1974).

11. A. L. Kamyshny, *Russ. J. Phys. Chem. 55*: 319 (1981).

12. L. K. James and L. D. Augenstein, *Adv. Enzymol. 28:* 1 (1966).

13. A. L. Kamyshny, in *Introduction to Applied Enzymology* (Russ.) (I. V. Berezin and K. Martinek, eds.), Moscow State University Publ. House, Moscow, 1982, pp. 62–100.

14. F. MacRitchie, *J. Colloid Interface Sci. 38*: 484 (1972).

15. F. MacRitchie, *Adv. Protein Chem. 32*: 283 (1978).

16. F. MacRitchie, *Colloids Surf. A 76*: 159 (1993).

17. C. A. Haynes and W. Norde, *Colloids Surf. B 2*: 517 (1994).

18. V. Lesins and E. Ruckenstein, *Colloid Polym. Sci. 266*: 1187 (1988).

19. W. Norde, *J. Dispersion Sci. Technol. 13:* 363 (1992).

20. W. Norde, *Clin. Mater. 11:* 85 (1992).

21. A. Kato and S. Nakai, *Biochim. Biophys. Acta 624*: 13 (1980).

22. A. F. H. Ward and L. Tordai, *J. Chem. Phys. 14*: 453 (1946).

23. A. F. H. Ward and L. Tordai, *Recueil 71*: 572 (1952).

24. F. MacRitchie, in *Proteins at Interfaces. Physicochemical and Biochemical Studies* (J. L. Brash and T. A. Horbett, eds.), ACS Symposium Series, Vol. 343, Washington, D.C., 1987, pp. 165–179.

25. R. Z. Guzman, R. G. Carbonell, and P. K. Kilpatrick, *J. Colloid Interface Sci. 114*: 536 (1986).
26. J. R. Hunter, P. K. Kilpatrick, and R. G. Carbonell, ibid. *142*: 429 (1991).
27. R. Douillard and J. Lefebvre, ibid. *139*: 488 (1990).
28. R. Douillard, J. Lefebvre, and V. Tran, *Colloids Surf. A 78:* 109 (1993).
29. R. Douillard, M. Daoud, J. Lefebvre, C. Minier, G. Lecannu, and J. Coutret, *J. Colloid Interface Sci. 163*: 277 (1994).
30. M. Blank, B. B. Lee, and J. S. Britten, ibid., *50*: 215 (1975).
31. E. Tornberg, *J. Sci. Food Agric. 29*: 762 (1978).
32. E. Tornberg, *J. Colloid Interface Sci. 64*: 391 (1978).
33. J. M. G. Lankveld and J. Lyklema, ibid. *41*: 454 (1972).
34. D. E. Graham and M. C. Phillips, ibid. *70*: 403 (1979).
35. J. A. De Feijter and J. Benjamins, in *Food Emulsions and Forms* (E. Dickinson, ed.), Royal Society of Chemistry, London, 1987, pp. 72–85.
36. D. E. Graham and M. C. Phillips, *J. Colloid Interface Sci. 70*: 415 (1979).
37. H. B. Bull, ibid. *41:* 305 (1972).
38. F. MacRitchie, *J. Macromol. Sci. A4*: 1169 (1970).
39. G. Gonzalez and F. MacRitchie, *J. Colloid Interface Sci. 105*: 55 (1970).
40. F. MacRitchie, ibid. *105*: 119 (1985).
41. M. C. Phillips, M. T. A. Evans, D. E. Graham, and D. Oldani, *Colloid Polym. Sci. 253*: 424 (1975).
42. D. J. Adams, M. T. A. Evans, J. R. Mitchell, M. C. Phillips, and P. M. Rees, *J. Polym. Sci. C34*: 167 (1971).
43. D. E. Graham and M. C. Phillips, *J. Colloid Interface Sci. 70*: 427 (1979).
44. A. Khaiat and I. R. Miller, *Biochim. Biophys. Acta 183*: 309 (1969).
45. J. R. Hunter, R. G. Carbonell, and P. K. Kilpatrick, *J. Colloid Interface Sci. 143*: 37 (1991).
46. C. C. Bigelow, *J. Theoret. Biol. 16*: 187 (1967).
47. E. Keshavarz and S. Nakai. *Biochim. Biophys. Acta 576*: 269 (1979).
48. C.-H. Lee and S.-K. Kim. *Food Hydrocolloids 1*: 283 (1987).
49. S. Damodaran and K. B. Song, in *Interactions of Food Proteins* (N. Parris and R. Barford, eds.), ACS Symposium Series, Vol. 454, Washington, D.C., 1991, pp. 104–121.
50. S. Magdassi, O. Sheinberg, and Z. Zakay-Rones, ACS Symposium on Proteins at Interfaces, in press.

51. J. E. Kinsella and D. M. Whitehead, in *Proteins at Interfaces. Physicochemical and Biochemical Studies* (J. L. Brash and T. A. Horbett, eds.), ACS Symposium Series, Vol. 343, Washington, D.C., 1987, pp. 629–646.

52. M. T. A. Evans, J. Mitchell, P. R. Mussellwhite, and L. Irons, *Adv. Exper. Med. Biol. 7*: 1 (1970).

53. S. H. Kim and J. E. Kinsella, *J. Food Sci. 50*: 1526 (1985).

54. R. D. Waniska and J. E. Kinsella, *J. Agric. Food Chem. 33*: 1143 (1985).

55. E. Tornberg, in *Functionality and Protein Structure* (A. Pour-El, ed.), ACS Symposium Series, Vol. 92, Washington, D.C., 1979, pp. 105–123.

56. G. Baum and M. Lynn, *Process Biochem. 10*: 14 (1975).

57. R. A. Messing, *J. Non-Cryst. Solids 19*: 277 (1975).

58. A. T. Kudish and F. R. Eirich, in *Proteins at Interfaces. Physicochemical and Biochemical Studies* (J. L. Brash and T. A. Horbett, eds.), ACS Symposium Series, Vol. 343, Washington, D.C., 1987, pp. 261–277.

59. T. A. Horbett, ibid., pp. 239–260.

60. J. L. Brash, ibid., pp. 491–505.

61. A. Sadana, *Chem. Rev. 92*: 1799 (1992).

62. A. Kondo, S. Oku, and K. Higashitani, *J. Colloid Interface Sci. 143*: 214 (1991).

63. A. Kondo and K. Higashitani, ibid. *150*: 344 (1992).

64. B. L. Steadman, K. C. Thompson, R. C. Middaugh, K. Matsuno, S. Vrona, E. Q. Lawson, and R. V. Lewis, *Biotechnol. Bioeng. 40*: 8 (1992).

65. D. Sarkar and D. K. Chattoraj, *J. Colloid Interface Sci. 157*: 219 (1993).

66. D. Sarkar and D. K. Chattoraj, *Colloids Surf. B2*: 411 (1994).

67. H. W. Morgan and C. T. Corke, *Canad. J. Microbiol. 22*: 684 (1976).

68. E. Glueckauf and L. Patterson, *Biochim. Biophys. Acta 351*: 57 (1974).

69. D. J. Hay and E. C. Moreno, *J. Dent. Res. 58*: 930 (1979).

70. W. Norde, J. G. E. M. Fraaye, and J. Lyklema, in *Proteins at Interfaces. Physicochemical and Biochemical Studies* (J. L. Brash and T. A. Horbett, eds.), ACS Symposium Series, Vol. 343, Washington, D.C., 1987, pp. 36–47.

71. W. Norde, *Pure Appl. Chem. 66*: 491 (1994).

72. R. A. Messing, *Methods Enzymol. 44*: 148 (1976).

73. H. H. Weetall, *Chimia 30*: 429 (1976).

74. P. Greenwald, W. Gunsser, and S. Scheer, *Angew. Chem. 89*: 761 (1977).
75. B. Solomon and Y. Levin, *Biotechnol. Bioeng. 17*: 1323 (1975).
76. R. Kurrat, J. J. Ramsden, and J. E. Prenosil, *J. Chem. Soc. Faraday Trans. 90*: 587 (1994).
77. L. Goldstein and G. Manecke, *Appl. Biochem. Bioeng. 1*: 23 (1976).
78. Y. K. Cho and J. R. Bailey, *Biotechnol. Bioeng. 20*: 1651 (1978).
79. W. J. Dillman and J. F. Miller, *J. Colloid Interface Sci. 44*: 221 (1973).
80. B. R. Young, W. G. Pitt, and S. L. Cooper, ibid. *124*: 28 (1988).
81. A. Kondo, F. Murakami, and K. Higashitani, *Biotechnol. Bioeng. 40*: 889 (1992).
82. D. V. Brooksbank, C. M. Davidson, D. S. Horne, and J. Leaver, *J. Chem. Soc. Faraday Trans. 89*: 3419 (1993).
83. H. Yoshida, H. Nishihara, and T. Kataoka, *Biotechnol. Bioeng. 41*: 280 (1993).
84. R. D. Tilton, C. R. Robertson, and A. P. Gast, *Langmuir 7*: 2710 (1991).
85. P. Van Dulm and W. Norde, *J. Colloid Interface Sci. 91*: 248 (1983).
86. G. Penners, Z. Priel, and A. Silberg, ibid. *80*: 437 (1981).
87. Y. L. Cheng, S. A. Darst, and C. R. Robertson, ibid. *118*: 212 (1987).
88. M. C. Wahlgren, T. Arnebrant, and M. Paulsson, ibid. *158*: 46 (1993).
89. A. C. Olson and W. L. Stanly, *Enzyme Eng. 2*: 91 (1974).
90. S. Magdassi, D. Leibler, and S. Braun, *Langmuir 6*: 376 (1990).
91. P. J. Yon and R. J. Simmond, *Biochem. J. 177*: 417 (1979).
92. A. L. Nikolaev, E. M. Benko, and G. D. Vovchenko, *Russ. J. Phys. Chem. 47*: 743 (1973).
93. M. A. Borisova, L. I. Nekrasov, N. I. Kobosev, A. V. Kiselev, and Yu. S. Nikitin, ibid. *43*: 291 (1969).
94. W. Norde and E. Rouwendal, *J. Colloid Interface Sci. 139*: 169 (1990).
95. H. Shirahama, J. Lyklema, and W. Norde, ibid. *139*: 177 (1990).
96. H. Sato, T. Tomiyama, H. Moritomo, and A. Nakajima, in *Proteins at Interfaces. Physicochemical and Biochemical Studies* (J. L. Brash and T. A. Horbett, eds.), ACS Symposium Series, Vol. 343, Washington, D.C., 1987, pp. 76–87.
97. E. Niemann and W. Hoffmann, *Z. Analyt. Chem. 200*: 443 (1964).
98. J. Lundstrom, *Progr. Colloid Polym. Sci. 70*: 76 (1985).

99. S. P. Mitra and D. K. Chattoraj, *Indian J. Biochem. Biophys. 15*: 147 (1978).
100. F. Kozin and D. J. McCarty, *J. Lab. Clin. Med. 89*: 1314 (1977).
101. B. D. Fair and A. M. Jamieson, *J. Colloid Interface Sci. 77*: 525 (1980).
102. M. E. Soderquist and A. G. Walton, ibid. *75*: 386 (1980).
103. J. Lyklema, *Colloids Surf. 10*: 33 (1984).
104. W. H. Grant, B. W. Morrissey, and R. R. Stromberg, *Amer. Chem. Soc. Polymer Prepr. 16*: 163 (1975).
105. A. N. Mitrofanova, E. S. Chukhrai, and O. M. Poltorak, *Russ. J. Phys. Chem. 49*: 1876 (1975).
106. M. N. Veselova, E. S. Chukhrai, and O. M. Poltorak, ibid. *48*: 1192 (1974).
107. W. Schöpp, J. Thifronitou, and H. Aurich, *Acta Biol. et Med. Ger. 35*: 1443 (1976).
108. H. L. Brockman, J. H. Law, and F. J. Kezdy, *J. Biol. Chem. 248*: 4965 (1973).
109. J. L. Brash and Q. M. Samak, *J. Colloid Interface Sci. 65*: 495 (1978).
110. P. K. Weathersby, T. A. Horbett, and A. S. Hoffman, *J. Bioeng. 1*: 395 (1977).
111. B. M. C. Chan and J. L. Brash, *J. Colloid Interface Sci. 82*: 217 (1981).
112. B. K. Lok, Y. L. Cheng, and C. R. Robertson, ibid. *91*: 104 (1983).
113. J. L. Bohnert and T. A. Horbett, ibid. *111*: 363 (1986).
114. P. K. Weathersby, T. A. Horbett, and A. S. Hoffman, *Trans. Am. Soc. Artif. Int. Organs 22*: 242 (1976).
115. F. Grinell and M. K. Feld, *J. Biomed. Mater. Res. 15*: 363 (1981).
116. D. F. Waugh, J. A. Lippe, and Y. R. Freund, ibid. *12*: 599 (1978).
117. D. F. Waugh, L. J. Anthony, and H. Ng, ibid. *9*: 511 (1975).
118. W. Norde and J. Lyklema, *J. Colloid Interface Sci. 66*: 295 (1978).
119. T. Arai and W. Norde, *Colloids Surf. 51*: 1 (1990).
120. W. Norde and J. P. Favier, ibid. *64*: 87 (1992).
121. B. M. C. Chan and J. L. Brash, *J. Colloid Interface Sci. 84*: 263 (1981).
122. B. W. Morrissey and C. A. Fenstermaker, *Trans. Amer. Soc. Artif. Int. Organs 22*: 278 (1976).
123. G. I. Loeb, *J. Colloid Interface Sci. 31*: 572 (1969).
124. V. Krisdhasima, J. McGuire, and R. Sproull, ibid. *154*: 337 (1992).
125. Y.-S. Lin and V. Hlady, *Colloids Surf. B 2*: 481 (1994).

126. H. Elwing, S. Welin, A. Askendal, U. R. Nilsson, and J. Lundstrom, *J. Colloid Interface Sci. 119*: 203 (1987).
127. W. Norde, *Adv. Colloid Interface Sci. 25*: 267 (1986).
128. J. D. Andrade, V. L. Hlady, and R. A. van Wagenen, *Pure Appl. Chem. 56*: 1345 (1984).
129. P. Bagchi and S. M. Birnbaum, *J. Colloid Interface Sci. 83*: 460 (1981).
130. B. Matuszewska, W. Norde, and J. Lyklema, ibid. *84*: 403 (1981).
131. P. G. Koutsoukos, W. Norde, and J. Lyklema, ibid. *95*: 385 (1983).
132. A. V. Elgersma, R. L. J. Zsom, W. Norde, and J. Lyklema, ibid. *138*: 145 (1990).
133. R. G. Lee, C. Adamson, and S. W. Kim, *Thromb. Res. 4*: 485 (1974).
134. T. Arnebrant and T. Nylander, *J. Colloid Interface Sci. 111*: 529 (1986).
135. S. Uniyal and J. L. Brash, *Thromb. Haemostas. 47*: 285 (1982).
136. J. L. Brash and P. ten Hove, ibid. *51*: 326 (1984).
137. A. Prins, in *Advances in Food Emulsions and Foams* (E. Dickinson and G. Stainsby, eds.), Elsevier Applied Science, London, 1988, pp. 91–122.
138. D. E. Graham and M. C. Phillips, in *Theory and Practice of Emulsion Technology* (A. L. Smith, ed.), Academic Press, London, 1976, pp. 75–98.
139. E. Tornberg and N. Ediriweera, in *Food Emulsions and Foams* (E. Dickinson, ed.), Royal Society of Chemistry, London, 1987, pp. 52–63.
140. E. Dickinson and G. Stainsby, in *Advances in Food Emulsions and Foams* (E. Dickinson and G. Stainsby, eds.), Elsevier Applied Science, London, 1988, pp. 1–44.
141. P. J. Halling, *CRC Crit. Rev. Food Sci Nutr. 15*: 155 (1981).
142. G. Stainsby, in *Functional Properties of Food Macromolecules* (J. R. Mitchell and D. A. Ledward, eds.), Elsevier Applied Science, London, 1986, pp. 315–353.
143. J. P. Cherry, K. H. McWatters, and L. R. Beuchat, in *Functionality and Protein Structure* (A. Pour-El, ed.), ACS Symposium Series, Vol. 92, Washington, D.C., 1979, pp. 1–26.
144. J. E. Kinsella and K. J. Shetty, ibid. pp. 37–63.
145. K. H. McWatters and J. P. Cherry, in *Protein Functionality in Foods* (J. P. Cherry, ed.), ACS Symposium Series, Vol. 147, Washington, D.C., 1981, pp. 217–242.

146. R. D. Phillips and L. R. Beuchat, ibid., pp. 275–298.
147. J. E. Kinsella and D. M. Whitehead, in *Advances in Food Emulsions and Foams* (E. Dickinson and G. Stainsby, eds.), Elsevier Applied Science, 1988, 163–188.
148. S. Arai and M. Watanabe, ibid., pp. 189–220.
149. A. Kato, H. Yamaoka, N. Mutsudomi, and K. Kobayashi, *J. Agric. Food Chem. 34*: 370 (1986).
150. R. D. Waniska, J. K. Shetty, and J. E. Kinsella ibid. *29*: 826 (1981).
151. R. H. Schmidt, V. Packard, and H. Morris *J. Dairy Sci. 67*: 2723 (1984).
152. J. B. German, T. E. O'Neill, and J. E. Kinsella, *J. Amer. Oil Chem. Soc. 62*: 1358 (1985).
153. S. Poole and J. Fry, in *Development in Food Proteins-5* (B. Hudson, ed.), Elsevier Applied Science, New York, 1987, pp. 257–298.
154. H. Hedenskog and H. Morgen, *Biotechnol. Bioeng. 15*: 129 (1973).
155. S. Magdassi and A. Stawsky, *J. Disp. Sci. Technol. 10*: 213 (1989).
156. S. Magdassi, A. Stawsky, and S. Braun, *Tenside 28*: 264 (1991).
157. R. Nakamura, S. Hayakawa, K. Yasuda, and Y. Sato, *J. Food Sci. 49*: 102 (1984).
158. K. M. Franzen and J. E. Kinsella, *J. Agric. Food Chem. 24*: 788 (1976).
159. K. N. Pearce and J. E. Kinsella, ibid. *26*: 716 (1978).
160. S. L. Woo and T. J. Richardson, *J. Dairy Sci. 66*: 984 (1983).
161. Y. T. Huang and J. E. Kinsella, *J. Food Sci. 52*: 1684 (1987).
162. R. D. Waniska and J. E. Kinsella, *Int. J. Peptide Protein Res. 23*: 467 (1984).
163. G. Puski, *Cereal Chem. 52*: 655 (1975).
164. J. Spinelli, B. Koury, and R. Miller, *J. Food Sci. 37*: 604 (1972).
165. C. A. Kuehler and C. M. Stine, ibid. *39*: 379 (1974).
166. J. Adler-Nissen, *J. Proc. Biochem. 12*: 18 (1977).
167. B. A. Lewis and J. H. Chen, in *Functionality and Protein Structure* (A. Pour-El, ed.), ACS Symposium Series, Vol. 92, Washington, D.C., 1979, pp. 27–35.
168. A. A. Sekul, C. H. Vinnett, and R. L. Ory, *J. Agric. Food Chem. 26*: 855 (1978).
169. A.-M. Hermansson, D. Olsson, and B. Holmberg, *Lebens.-Wiss. u. Technol. 7*: 176 (1974).
170. L. P. Grunden, D. V. Vadehra, and R. C. Baker, *J. Food Sci. 39*: 841 (1974).
171. J. D. Ferry, *Adv. Protein Chem. 4*: 1 (1948).

172. E. Barbu and M. Joly, *Faraday Discuss. Chem. Soc. 13*: 77 (1953).
173. P. Kratchovil, P. Munk, and B. Sedlacek, *Coll. Czech. Chem. Commun. 27*: 115 (1962).
174. P. Kratchovil, P. Munk, and B. Sedlacek, ibid. *27*: 788 (1962).
175. N. K. Howell, in *Developments in Food Proteins*, Vol. 7 (B. J. F. Hudson, ed.), Elsevier Applied Science, London, 1991, pp. 231–270.
176. M. G. Bezrukov, *Angew. Chem. 18*: 599 (1979).
177. F. Zirbel and J. Kinsella, *Milchwissenschaft 43*: 691 (1988).
178. M. E. Mangino, *J. Dairy Sci. 67*: 2711 (1984).
179. R. H. Schmidt, B. L. Illingworth, and E. M. Ahmed, *J. Food Sci. 43*: 613 (1978).
180. R. H. Schmidt, B. L. Illingworth, and E. M. Ahmed, *J. Food Proc. Pres. 2*: 111 (1978).
181. K. Shimada and S. Matsushita, *J. Agric. Food Chem. 28*: 409 (1980).
182. K. Shimada and S. Matsushita, ibid. *28*: 413 (1980).
183. S. Utsumi and J. E. Kinsella, *J. Food Sci. 50*: 1278 (1985).
184. M. S. B. Joseph and M. E. Mangino, *Austr. J. Dairy Technol. 5*: 9 (1988).
185. L. P. Voutsinas, E. Cheung, and S. Nakai, *J. Food Sci. 48*: 26 (1983).
186. D. M. Mulvihill and J. E. Kinsella, ibid. *53*: 231 (1988).
187. J. E. M. Johns and B. M. Ennis, *New Zeal. J. Dairy Sci. Technol. 15*: 79 (1981).
188. S. Hayakawa and S. Nakai, *Can. Inst. Food Sci. Technol. J. 18*: 290 (1985).
189. K. Yasuda, R. Nakamura, and S. Hayakawa, *J. Food Sci. 51*: 1289 (1986).

2
Enhanced Hydrophobicity
Formation and Properties
of Surface-Active Proteins

Shlomo Magdassi and Ofer Toledano

The Hebrew University of Jerusalem, Jerusalem, Israel

I. INTRODUCTION

Since proteins are composed of both hydrophobic and hydrophilic amino acids, most of them have an amphipathic nature. For example, tryptophan is considered as a hydrophobic amino acid, while serin is considered as a hydrophilic amino acid. The overall protein hydrophobicity, which can be calculated, combines the effects of the peptide sequence of the backbone and the type of residues buried inside the core of the protein molecule or the residues present in the outer surface of the molecule.

As described in Chapter 1, various factors may affect the surface activity of proteins, hydrophobicity being a dominant parameter.

In nature, bovine serum albumin has a unique surface activity as compared to lysozime. The former is much more hydrophobic, having a relative hydrophobicity index (based on the fluorescence method) of 100, compared to 7 for lysozyme [1,2].

The importance of the hydrophobicity of the protein in determining the surface activity was demonstrated clearly by Kato and Nakai [2]. As shown in Fig. 1a, the emulsifying activity index increases with the increase in the hydrophobicity index of various proteins. In a similar way, the increase in hydrophobicity index leads to a decrease in the interfacial tension (water–corn oil) (Fig. 1b). Although these

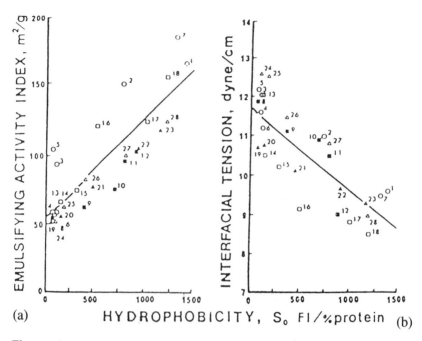

(a)   HYDROPHOBICITY, S_0 FI/%protein (b)

Figure 1 Relationships of S_0 with interfacial tension and emulsifying activity of proteins. 1, bovine serum albumin; 2, β-lactoglobulin; 3, trypsin; 4, ovalbumin; 5, conalbumin; 6, lysozyme; 7, κ-casein; 8, 9, 10, 11, and 12, denatured ovalbumin by heating at 85°C for 1, 2, 3, 4, and 5 min respectively; 13, 14, 15, 16, 17, and 18, denatured lysozyme by heating at 85°C for 1, 2, 3, 4, 5, and 6 min respectively; 19, 20, 21, 22, and 23, ovalbumin bound with 0.2, 0.3, 1.7, 5.7, and 7.9 mol of sodium dodecyl sulfate/mol of protein respectively; 24, 25, 26, 27, and 28, ovalbumin bound with 0.3, 0.9, 3.1, 4.8, and 8.2 mol of linoleate/mol of protein respectively. Interfacial tension: measured at corn oil/0.2% protein interface with a Fisher Surface Tensiomat Model 21. Emulsifying activity index: calculated from the absorbance at 500 nm of the supernatant after centrifuging blended mixtures of 2 ml of corn oil and 6 ml of 0.5% protein in 0.01 M phosphate buffer, pH 7.4 S_0: initial slope of fluorescence intensity (FI) vs. percent protein plot. 10 μl of 3.6 mM *cis*-parinaric acid solution was added to 2 ml of 0.002 to 0.1% protein in 0.01 M phosphate buffer, pH 7.4, containing 0.002% SDS. FI was measured at 420 nm by exciting at 325 nm. (From Ref. 2. Reprinted by permission.)

correlations are impressive, it should be emphasized that other important parameters, such as the tertiary structure, disulphide bridges, etc., are also very important.

The contribution of amino acids to the overall hydrophobicity of the protein is limited, compared to the contribution of a hydrophobic tail in classic surfactants, such as ethoxylated fatty acids. Therefore it can be expected that incorporation of long hydrocarbon chains will affect significantly the surface activity of the protein, as will be described in the following sections.

II. EVALUATION OF PROTEIN HYDROPHOBICITY

The common way to determine the hydrophobicity of a given molecule is based on its solubility in a polar or nonpolar medium. As most proteins are not soluble in nonpolar solvents, there was a need to use other methods to describe quantitatively the hydrophobicity of proteins. Because there is no standard method, we shall describe briefly some of the most common ones [4].

A. Fluorescence Methods

These methods are based on evaluation of the shift of peak fluorescence wavelengths, which are dependent on the polarity of the environment inside and at the protein surface, and they utilize either the fluorescence of various amino acids (such as tryptophan) [5] present in the protein or that of various hydrophobic probes such as 1-anilinonaphthalene-8-sulfonate (ANS) [6–10]. Various quenching additives may also supply information, particularly on the surface hydrophobicity.

B. Partition Coefficients

By these methods, the protein is partitioned in a system containing two phases, each phase having a different polarity. From the concentrations of protein in each phase, the energy needed to transfer a protein molecule from one phase to the other can be calculated and used as a hydrophobic parameter value. Such systems are based on special combinations of polymers, such as dextran and polyethylene glycol, dissolved in aqueous solutions [11,12]. Similarly, hydrophobic chromatography can be used, in which the retention time serves as a measure guide for hydrophobicity [13–19].

C. Hydrophobic Binding

This method is based on the binding of hydrophobic markers to the protein molecule, while the adsorbed quantity is used as a measure for hydrophobicity. This molecule can be a simple aliphatic molecule such as hexan [20–25] or triglyceride [26,27] or even a surfactant such as sodium dodecyl sulphate [28–30].

It should be emphasized that all these methods give only relative hydrophobicity values. In contrast, for the hydrophobically modified proteins, it is possible to determine the exact number of hydrophobic groups attached to each protein molecule. Such methods are based on the determination of the amount of attached groups by radioactively labeled groups [31], elementary analysis [63], or even sophisticated mass spectrometry methods [33]. The degree of modification can also be determined by measuring the number of unreacted modification sites on the protein, before and after modification.

For example, if the modification reaction takes place via the lysine groups, the free lysine groups can be determined by ninhydrin [32,34], trinitrobenzenesulphonic acid [35,37], ortho pthalaldehyde [38,62], etc.

III. HYDROPHOBIC MODIFICATIONS

Increased hydrophobicity of proteins can be achieved either by changing the structure or the molecular weight, or by covalent attachment of various hydrophobic groups to the protein molecule. In addition, attachment of hydrophobic groups can be obtained by adsorption of various molecules on the protein. For example, electrical interactions between ionic surfactants and the charged parts of the protein may lead to formation of protein–surfactant complexes, which have various degrees of hydrophobicity. Such a case is the binding of sodium dodecyl sulphate with a positively charged gelatin molecule [49,40]. The binding ratio can vary according to the initial molar ratio of the two components, making this modification method very simple. However, this kind of modification might be reversible, and the degree of modification is strongly dependent on solution conditions, dilution, etc. These kinds of modifications will be discussed in more detail in the chapter by Jones, which deals with surfactant–protein interactions.

A. Protein Denaturation

In general, the hydrophobic portions of the amino acids tend not to be in contact with the aqueous phase and are buried in the interior of the protein molecule. Denaturation, which is discussed in a separate chapter, is a process in which the protein loses its tertiary structure, which may lead, in many cases, to exposure of the hydrophobic parts towards the aqueous phase. The end result is an increased hydrophobicity of the protein surface, which affects significantly its surface activity. As shown by Graham and Phillips, heating lysozyme (above 50°C, pH = 1), caused a dramatic change in the surface activity at air–water interface [41]. Heating ovalbumin for only 3 min at 80°C (pH = 5.6) caused an increased hydrophobicity of the protein accompanied by a tenfold increase in its emulsifying stability index and about a twofold increase in its emulsifying activity index. These kinds of observations were reported by Voutsinas et al. [42] for various proteins. It is obvious that denaturation, which may also be achieved by specific chemical reactions such as breaking of disulphide bonds [43], would always lead only to such a degree of hydrophobicity as is limited by the primary amino acid composition of the native protein.

B. Covalent Attachment of Hydrophobic Groups

It is clear that covalent attachment of hydrophobic groups to the protein molecule, in order to increase its hydrophobicity, has many advantages over denaturation and physicochemical binding. By using various modifying methods one can control hydrophobicity by changing the degree of modification (the number of attached groups per protein molecule) or the length of each attached chain, or even the type of hydrophobic groups. Controlling the site of modification may also contribute to the overall hydrophobicity of the protein molecule: for example, amidation of free lysine groups present in the protein, by a fatty acid, reduces the number of positive sites present in the protein, at pH range below the pKa of the amino group. This combined effects of increasing the hydrophobic moieties and decreasing the hydrophilicity affect significantly the surface activity of the protein, as will be shown for various covalent attachment methods.

Furthermore, the attachment of the hydrophobic tails will affect the surface activity more than by simply increasing hydrophobicity. As presented in Fig. 2, the attachment of hydrophobic tails will cause several domains of the protein to behave as a classical surfactant, and the overall effect will be the formation of a polymer-like surfactant. Such a polymer, which will be adsorbed at various hydrophobic interfaces, should have an increased surface activity due to the existence of many adsorption sites per molecule. This adsorption mechanism may affect the reversibility of the adsorption process and might even prevent surface denaturation. This latter effect is important in appli-

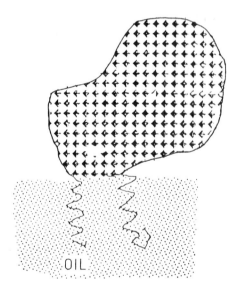

Figure 2 Schematic representation of protein molecule with chemically bound hydrophobic groups, at oil–water interface.

cations that combine surface and biological activity, such as enzyme immobilization.

Covalent attachment of various hydrophobic groups can be achieved by numerous reactions. Such reactions can be found in the books by Means and Feeny [44] and Lundblad and Noyes [45]. In this chapter we shall describe only some modification methods, mainly those in which the resulting modified proteins were also evaluated in terms of surface activity. It should be emphasized that the reaction type should be chosen individually for each protein, according to various factors, such as solubility in various solvents, denaturation of the protein, etc.

1. Acylation

The most common covalent attachments are acylation reactions, in particular on the lysine groups. Such attachment can be performed by the following reactions:

(a) *Anhydrides.* Anhydrides of various carboxylic acids can react with lysine groups, as shown in the scheme

$$\text{(P)}-\text{NH}_2 + \begin{matrix} \text{O} \\ \| \\ \text{C}-\text{R} \\ \diagdown \\ \diagup \\ \text{C}-\text{R} \\ \| \\ \text{O} \end{matrix} \longrightarrow \text{(P)}-\text{NH}-\overset{\overset{\text{O}}{\|}}{\text{C}}-\text{R} + \text{RCOO}^- + \text{H}^+$$

in which R may vary from CH_3 to $C_{17}H_{35}$. The most common reagent is acetic anhydride, which was used to modify various proteins such as yeast protein [46], casein [47,48], whey protein [49], fish protein [50–52], soy protein [53], cottonseed protein [54], wheat flower protein [55], and egg white protein [56–61].

Attachment of longer hydrophobic chains via anhydrides were reported by several investigators: Evans et al. have covalently linked β-casein with a series of hydrophobic groups up to decanoyl [79], and Arranz et al. have linked stearoyl groups to gelatin, up to 3.8×10^{-4} eq/g gelatin [63].

The degree of acylation can be controlled by simply changing the molar ratio of the anhydride to protein, leading to proteins that have

Table 1 Effect of Acetylation on Isoionic pH, Relative Mobility, and Extinction Coefficient of Ovalbumin

Molar ratio of anhydride to protein	ε-NH₂ groups modified (%)	Isoionic point	Relative mobility	Molar extinction coefficient at 280 nm, × 10⁻³
0	0	4.9	0.28	30.1
11	21	4.6	0.33	29.5
22	28	4.5	0.34	29.0
88	62	4.2	0.37	27.9
110	66	4.2	0.37	27.2
880	93	4.1	0.54	21.2
1100	98	4.1	0.55	20.3

Source: Ref. 64.

various degrees of hydrophobicity and thus various surface activities. As demonstrated in Table 1, increasing the molar ratio up to 1100 acetic anhydride molecules per ovalbumin molecule leads to attachment of 18.6 acetyl groups [64]. For gelatin, 118 stearic anhydride molecules lead to attachment of 38 groups onto a gelatin molecule. These types of reactions may be performed in aqueous solutions [64] or in organic solvents such as DMSO [63]. It should be noted that the anhydrides may hydrolyze rapidly in the presence of water and that therefore such modifications are often performed by having a higher anhydride molar ratio to the lysine groups content, and by addition of the reagent in aliqoats. Moreover, the solubility of long-chain anhydrides is very low at aqueous solutions, and these reactions may require additional solubilizing agent. On the other hand, performing such reactions in organic, dry solvent should lead to higher yields for long-chain anhydrides too. However, the main problem here is to dissolve the protein in the organic solvent. In such cases, special care should also be taken to prevent formation of highly modified proteins that would be further insoluble in water, as were obtained for modified gelatin by stearic groups [63]. While using anhydrides, O-acylation may also occur, although to a lesser extent, as demonstrated by Ansari et al. [64], for modification of ovalbumin by acetic anhydride.

(b) *Acyl Chloride.* Various acyl halides may react with lysine groups as shown in the scheme

$$\text{P}-NH_2 + \underset{\underset{O}{\overset{\|}{}}}{\overset{X}{\underset{}{C}}}-R \longrightarrow \text{P}-NH-\underset{\underset{}{\overset{O}{\overset{\|}{C}}}}{}-R + HX$$

This type of reaction was performed by Torchilin et al. [65] by reacting the enzyme α-chymotrypsin with palmitic acyl chloride. They could covalently attach between one and six palmitoyl residues to an enzyme molecule, with very little effect on the enzymatic activity. These modified enzymes were further incorporated into the membranes of liposomes, via the hydrophobic tail, while the protein molecule was located on the liposome membrane.

Such reactions were also reported for commercial cosmetic products [66], in which hydrolized proteins were reacted with various fatty acid chlorides to yield protein–fatty acid condensates that have detergent properties.

(c) N-*Hydroxy Succinimide Esters.* By using various active esters, it is possible to attach hydrophobic groups to the protein's lysine groups only, according to the scheme

$$\text{P}-NH_2 + R-\overset{\overset{O}{\|}}{C}-O-N \longrightarrow \text{P}-NH-\overset{\overset{O}{\|}}{C}-R$$

$$+ \ H-O-N$$

The *N*-hydroxy succinimide is a good leaving group, causing such esters to be very active in mild conditions. Such modifications led to various palmitoyl–soy protein derivatives [67,73] without destroying the globular structure of the protein. Magdassi et al. have modified

ovalbumin with various hydrophobic groups [68–70] and showed increased surface activity upon modification, as will be seen later. This
reaction is also suitable when a specific biological activity of the protein should be retained, due to mild reaction conditions at aqueous
solutions. As demonstrated for modified IgG by Huang et al. [71,72],
attachment of several C_{16} groups did not reduce the biological recognition ability of the antibodies. While the modified antibodies were
attached to liposomes through the hydrophobic tails, a new drug targeting system was established.

Obviously, this reaction can also be performed in organic solvents
in order to prevent hydrolysis of the active esters.

2. Enzymatic Modifications

Increased hydrophobicity may be achieved by enzymatic hydrolysis,
which can lead to rearrangements of the buried hydrophobic groups
present in the native globular protein [74]. On the other hand, the
opposite effect may be observed if the end products are low molecular weight segments [75]. Therefore an optimum in surface activity
is achieved at a specific molecular weight range for each protein [75].

Incorporation of various hydrophobic residues was reported by
Arai and Yamashita [76], using aminolysis through the plastein reaction. As shown schematically in Fig. 3, L-leucine n-dodecyl ester
was covalently bound to succinylated α-casein by using papain. This
method allows the incorporation of the hydrophobic groups to specific amino acids present in the protein backbone. Such procedures
were carried out also for gelatin and soy proteins. The resulting modified proteins had various molecular weights, (2000–40,000 daltons),
and the hydrophobic chains varied from C_2 to C_{12}. It was found that
some of these modified proteins were more surface active than the
native proteins [77]. It also appears that the enzymatic modifications
are more limited than the chemical modifications. However, it is very
likely that such enzymatic modifications would have an advantage in
cases where retention of biological activity is desired, as in possible
enzyme immobilization. These advantages, together with various enzymatic modifications, are discussed in more detail in the chapter by
G. Hajós.

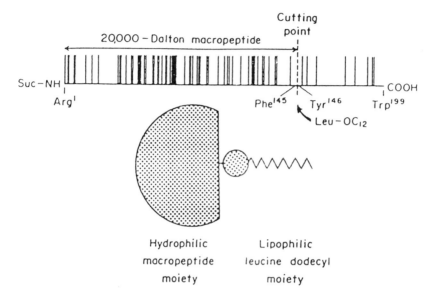

Figure 3 Covalent incorporation of L-leucine n-dodecyl ester into succinylated α_{s1} casein by modification with papain and the resulting formation of a 20,000-dalton macropeptide having an amphiphilic structure. The position of hydrophilic amino acid residues on the protein molecule are indicated with vertical bars. (From Ref. 3. Reprinted by permission.)

IV. SURFACE ACTIVITY OF HYDROPHOBICALLY MODIFIED PROTEINS

The large variety of hydrophobic groups that might be incorporated into the protein molecule opens a very wide range of possibilities to obtain improved surface activity. As mentioned earlier, it is possible to change both the number and the chain length of the hydrophobic moieties. These lead to formation of amphiphilic proteins, which resemble polymeric surfactants. In this section we describe the effect of attachment of various hydrophobic groups on the surface activity evaluated in various interfaces and dispersed systems.

A. Surface and Interfacial Tension

Haque and Kito [78] attached several fatty acid residues to a α_{s1}-casein and found that the surface tension was reduced with increase in chain length up to a limiting chain length from which no further reduction was observed. For example, the surface tensions of the C_8, C_{12}, C_{16}, and C_{18} are 54.2, 45.9, 38.9 and 37.4 dyne/cm respectively. Introduction of double bonds in the hydrophobic moiety does not have a significant effect on the surface tension reduction.

A similar observation was reported by Magdassi et al. [70] for modified ovalbumin. As shown in Fig. 4, attachment of C_2–C_{18} hydrophobic residues, at both high and low levels of degree of modification, causes a decrease in interfacial tension (water–tetradecane) while increasing chain length. It is interesting to note that the most significant decrease is already achieved at C_6 for the highly modified ovalbumin. Furthermore, it was found [70] that increasing the degree of modification while maintaining a constant chain length (C_{12}) improves the ability of the modified protein to reduce surface tension. It is not clear yet which parameter (chain length or degree of modification) is the most dominant.

Surface pressure measurements conducted in Langmuir film balance showed that at a constant area per molecule the attachment of

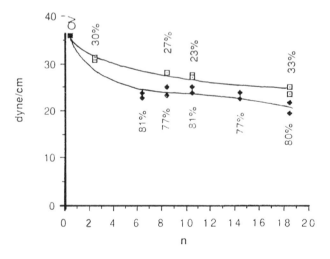

Figure 4 Effect of chain length of modified proteins on interfacial tension (water–tetradecan), at low and high degrees of modifications. The degree of modification of each protein is indicated. (From Ref. 70. Reprinted by permission.)

the longer chain length (C_{10}) leads to twice the surface pressure of that obtained by the native or the C-2 modified β-casein [79]. As expected, it was found that at a constant surface pressure the area per molecule increased with the increase in chain length.

Shimada et al. [80] showed that enzymatically modified gelatins displayed similar behavior, but it appears that the time needed to reach equilibrium surface tension is reduced significantly by the attachment of the hydrophobic groups.

Moreover, they have also shown that a C-12 modified gelatin reduced surface tension more efficiently than a conventional synthetic surfactant such as Tween 60 (ethoxylated sorbitan monostearate).

From the above results it is clear that reduction of surface tension may be achieved by modified proteins or by attachment of hydrophobic groups with or without further hydrolysis of the protein backbone.

B. Foaming

Adsorption of native and modified proteins at the air–water interface may lead to foam formation having various stabilities. As shown by Shimada et al. [80], foam stability is optimal at alkyl chain length of 6 in modified gelatin. The foam became completely unstable while the chain length exceeded C_{12}. Haque et al. [81] showed that the foaming activity was dependent on the number of the hydrophobic groups that were attached to α_{S1} casein, while the foam stability increased with the increase in the number of attached hydrophobic groups. It was suggested that the foam activity is related to the tendency for association of the surface-active protein [81].

These results prove that if suitable hydrophobic modifications are performed (chain length, degree of attachment), surface-active proteins that have better foaming properties than the native proteins can be achieved. The modified proteins can be used for various applications such as food foams (whipped cream) or cosmetic foam-forming additives.

C. Emulsification

The emulsification properties of the native and modified proteins can be evaluated by two parameters:

1. Emulsification activity (EA), which gives information about the ability of a protein to yield an emulsion at given conditions (protein concentration, pH, etc.), while increasing the oil–water ratio
2. Emulsion stability, which gives information about the ability of the protein to form a stable emulsion

Obviously these two parameters are not parallel, and one can obtain a small initial droplet size but an unstable emulsion.

Haque and Kito [81] modified α_{S1} casein by various degrees of attachment of palmitoyl groups. They found that at low soybean oil–water ratio there was no difference in emulsification activity for either native proteins or various other modified proteins. However, up to a 1:1 oil–water ratio, increasing the oil content led to a decrease in EA for the native protein. On the other hand, attachment of the hydrophobic groups led to an increase in EA. It is interesting to note that the improvement in emulsification activity can be observed already at a very low degree of modification—0.3 groups per 1 casein molecule—while the increase in modification of up to 11.1 groups did not change the EA significantly. A similar result was obtained for low and high modification degrees, for soybean protein [67]. The stability of the emulsion is more dependent on the degree of modification: for the palmitoyl derivatives of casein, unstable emulsions were obtained when the degree of modification was low (up to 2.5 groups per molecule). When the degree of modification was above this degree, stable emulsions were obtained.

Magdassi and Stavsky [69] have also observed that the emulsification properties have improved as a result of the attachment of various hydrophobic groups to ovalbumin. As compared to emulsions prepared by native ovalbumin [69], it was found that emulsification of tetradecane by any modified ovalbumin led to a significant increase in emulsion stability, together with a decrease in droplet size. The improvement in emulsion stability was most significant with highly modified proteins. For example, C_8 or C_6 modified ovalbumin (81% degree of modification) led to the formation of very stable emulsions, in which no oil separation or change in droplet size distribution was observed, even 50 days after the emulsion was prepared.

Increasing the degree of modification led to a decrease in the initial droplet size (which can be regarded as an emulsification activity index) for all chain lengths of the hydrophobic groups (C_2–C_{18}). For example, increasing the degree of modification from 5.4 to 15.4 hydrophobic groups per ovalbumin molecule led to a decrease in the value of median droplet size from 20 μm to 11 μm. This behavior could result from (a) an increase in the degree of modification, leading to a decrease in interfacial tension [70], and thus decreasing the droplet size and emulsion stability; (b) an increase in the number of anchoring chains of the protein at the oil–water interface, thus leading to better adsorption and coverage at the oil–water interface; (c) a higher modification degree, which practically reduces the positively charged group, causing the highly modified proteins to be more negatively charged. Hence emulsion stability may increase due to electrical repulsion according to the DLVO theory.

The effect of chain length was evaluated for two groups of proteins having low and high modification degrees. As shown in Fig. 5, an optimal chain length of the hydrophobic groups is observed at C_8. This behavior, although less dramatic, was observed also for the proteins with low degrees of modification. Since the optimal chain

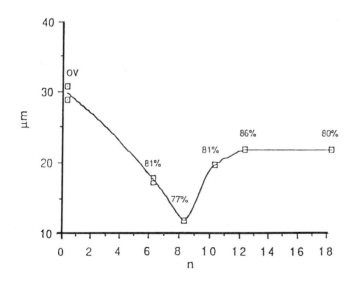

Figure 5 The effect of chain length of hydrophobic groups of the modified proteins (high modification degree) on median droplet size of the emulsions. (From Ref. 69. Reprinted by permission.)

length was observed for both high and low degrees of modification, it appears that the optimal chain length is independent of the overall hydrophobicity of the protein and is not correlated to the hydrophilic–lipophilic balance of the modified protein molecule. We assume that these results reflect the efficiency of the hydrophobic moieties to adsorb at the oil–water interface, on the one hand, and the tendency of the modified proteins to self aggregate via hydrophobic interactions, on the other hand.

It is, therefore, clear that hydrophobic modifications of proteins can affect their ability both to form and to stabilize oil in water emulsions. Special care should be taken in using the proper combination of chain length and number of attached groups to the protein molecule in order to obtain good protein-based emulsifiers.

D. Adsorption at Solid–Liquid Interfaces

Although a large number of studies have been conducted on adsorption of native proteins at solid–liquid interfaces, the modified proteins have not been studied systematically to the best of our knowledge. However, there are several commercial products that are based on condensation reactions between acyl chloride and hydrolyzed proteins, which are used for cosmetic hair products [82] and are obviously based on the ability of the modified proteins to adsorb at the hair–water interface. These products are based on electrostatic adsorption, which leads to the presence of hydrophobic moieties on the hair fibers, and hence a "conditioning" effect is observed.

Attachment of various hydrophobic groups to proteins usually increases the adsorbed amount onto solid particles. For example, attachment of stearyl groups, at various degrees of modification, leads to increased adsorption onto hydrophobic silica particles [68], as shown in Fig. 6. It is interesting to note that the various modified proteins obeyed the Langmuir adsorption model and that the adsorption energy, which was calculated according to this model, was not changed significantly upon modification. The general trend of increased adsorption with the increase in degree of modification was also observed for several other proteins, which have been tested in our laboratory but not reported yet. Among them are enzymes and

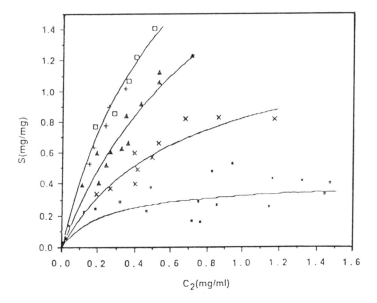

Figure 6 Adsorption isotherms of native and modified ovalbumins: ■, native ovalbumin; x, 13%; ▲, 24%; □, 33%; +, 57% modification. (From Ref. 68. Reprinted by permission.)

immunoglobulins, whose adsorption may lead to unique physico-chemical immobilization techniques while retaining biological activity.

V. BIOLOGICAL APPLICATIONS BASED ON HYDROPHOBICALLY MODIFIED PROTEINS

If a native protein retains its biological activity after hydrophobic modification, unique applications can be achieved. These applications are generally based on the enhancement of adsorption properties of the protein at various interfaces. For example, hydrophobic modifications of antibodies led to their increased adsorption onto oil droplets in O/W emulsions [83]. Therefore the emulsion droplets can be targeted via specific recognition (antibody–antigen) to specific sites, leading to the formation of drug targeting and diagnostic systems.

Huang et al. [71,72] used the attachment of hydrophobic groups to antibodies in the introduction of antibody molecules to liposomes, in which the hydrophobic tail is adsorbed into the bilayer or even penetrates it. These "immunoliposomes" could be further used for drug targeting.

Torchilin et al. used a similar approach to introduce enzymes into liposomes [65]. Formation of a surfactant–antibody conjugate as reported by Kabanov et al. [84] made it possible to link antibodies into micelles (which were formed by classical surfactants), which could solubilize hydrophobic probe molecules. These systems were suggested for drug delivery and targeting.

We have also recently found that upon attachment of various hydrophobic groups to antibody molecules, under certain conditions, the modified proteins have the ability to form spontaneous micelle-like structures [85], which in turn are able to solubilize hydrophobic molecules [86].

VI. SUMMARY

Enhancement of the surface activity of proteins can be achieved by hydrophobic modifications. These modifications are based on the exposure of hydrophobic groups to the outer part of the protein molecule, or by attachment of various groups to the protein molecule. These modifications generally lead to polymer-like surface-active molecules, which are more active at oil–water, air–water, solid–air, and solid–water interfaces. Unique applications, based on such modifications, can be achieved in cosmetics, pharmaceutics, food, and many other industries.

REFERENCES

1. A. Kato, T. Matsuda, N. Matsudomi, and K. Kobayashi, *J. Agric. Food Chem. 32*: 284 (1972).
2. A. Kato and S. Nakai, *Biochim. Biophys. Acta 624*: 13 (1980).
3. S. Arai and M. Watanabe, in *Properties of Water in Foods* (D. Simata and J. L. Multon, eds.), 1985, p. 49.
4. S. Nakai and E. Li-chan, in *Hydrophobic Interactions in Food systems,* Chap. 2, CRC Press, 1988.

5. E. A. Burstein, N. S. Vedenkina, and M. N. Ivkova, *Photochem. Photobiol. 18*: 263 (1973).
6. D. C. Turner and L. Brand, *Biochemistry 7*: 3381 (1968).
7. H. Terada, K. Hiramatsu, and K. Aoki, *Biochim. Biophys. Acta 622*: 161 (1980).
8. J. R. Daban, and M. D. Guasch, *Biochim. Biophys. Acta 625*: 237 (1980).
9. G. Palumbo and G. Ambrosio, *Arch. Biochem. Biophys. 212*: 37 (1981).
10. G. Penzer, *Eur. J. Biochem. 25*: 218 (1972).
11. V. P. Shankhay and C. C. Exelsson, *Eur. Biochem. 60*: 17 (1975).
12. B. Y. Zaslavsky, L. M. Miheeva, and S. V. Rogozhin, *Biochim. Biophys. Acta 588*: 89 (1979).
13. S. Hjerten, *J. Chromatog. 101*: 281 (1974).
14. W. Kissing and R. H. Reiner, *Chromatographia 10*: 129 (1977).
15. D. Yamashiro, *Int. J. Peptide Protein Res. 13*: 5 (1979).
16. M. J. O'Hare and E. C. Nice, *J. Chromatog. 171*: 209 (1979).
17. J. L. Meek, *Proc. Natl. Acad. Sci. U.S.A. 77*: 1632 (1980).
18. T. Imoto and K. Okazaki, *J. Biochem. 89*: 437 (1981).
19. K. J. Wilson, A. Honegger, R. P. Stotzel, and G. J. Hughes, *Biochem. J. 199*: 31 (1981).
20. H. G. Maier, *Zeit. Lebensm. U. Forsch. 141*: 332 (1970).
21. H. G. Maier, *Dtsch. Lebensm. -Rundschau 70*: 349 (1974).
22. H. G. Maier, *Proc. Int. Symp. Aroma Res. Zeist. 143* (1975).
23. C. C. Bigelow, *J. Theoret. Biol. 16*: 187 (1967).
24. A. Mohammadzadeh-K, R. E. Feeney, and L. M. Smith, *Biochim. Biophys. Acta 194*: 246 (1969).
25. A. Mohammadzadeh-K, L. M. Smith, and R. E. Feeney, *Biochim. Biophys. Acta 194*: 256 (1969).
26. L. M. Smith, P. Fantozzi, and R. K. Creveling, *J. Am. Oil Chem. Soc. 60*: 960 (1983).
27. T. Tsutsui, E. Li-Chan, and S. Nakai, *J. Food Sci. 51*: 1268 (1986).
28. J. A. Reynolds, and C. Tanford, *Proc. Natl. Acad. Sci. U.S.A. 66*: 1002 (1970).
29. O. Takenaka, S. Aizawa, Y. Tamura, J. Hirano, and Y. Inada, *Biochim. Biophys. Acta 263*: 696 (1972).
30. A. Kato, T. Matsuda, N. Matsudomi, and K. Kobayashi, *J. Agric. Food Chem. 32*: 284 (1984).
31. G. L. Peterson, *Methods in Enzymology 91*: 95 (1983).
32. S. Moor, *J. Biol. Chem. 243*: 6281 (1968).

33. M. Admczyk, A. Buko, Y. Chen, J. R. Fishpaugh, J. C. Gebler, and D. O. Johanson, *Bio. Conjugate Chem.* 5(6): 631 (1994).
34. H. Fraenkel-Cornat, *Methods in Enzymology 4*: 247 (1953).
35. J. Adler-Nissan, *J. Agric. Food Chem.* 27(6): 1256 (1979).
36. A. F. S. A. Hakeeb, *Anal. Biochem. 14*: 328 (1966).
37. K. Satke, T. Okazama, M. Owashi, and T. Shinoda, *J. Biochem.* 47(5): 654 (1960).
38. E. Mender, *Anal. Biochem. 127*: 55 (1982).
39. S. Magdassi and Y. Vinetsky, *J. Microencapsulation, 12*: 537 (1995).
40. V. Sarilj, L. Djakovec, and D. Dokic, *J. Colloid Interface Sci. 158*(2): 483 (1993).
41. D. E. Graham and M. C. Phillips, *J. Colloid Interface Sci.* 70(3): 427 (1979).
42. L. P. Voutsinas, E. Cheung, and S. Nakai, *J. Food Sci. 48*: 26 (1983).
43. R. M. Hillier, R. L. J. Lyster, and G. C. Cheeseman, *J. Sci. Food Agric. 31*: 1152 (1980).
44. G. E. Means and R. E. Feeny, in *Chemical Modification of Proteins*, Holder Day, 1971.
45. R. L. Lundblad and C. M. Noyes, in *Chemical Reagents for Protein Modification*, Vol. 1, CRC Press, 1985.
46. J. E. Kinsella, in *Food Proteins*, (P. F. Fox and J. J. Condon, eds.), Applied Science Publishers, London, 1982, p. 3.
47. M. Evans, L. Iron, and J. H. P. Petty, *Biochim. Biophys. Acta 243*: 259 (1971).
48. P. D. Hoagland, *Biochemistry 7*: 2542 (1968).
49. L. U. Thompson and E. S. Reyes, *J. Dairy Sci. 63*: 715 (1980).
50. L. Chen, T. Richardson, and C. Amundson, *J. Milk Food Technol. 38*: 89 (1975).
51. H. Groninger, Jr., and R. Miller, *J. Food Sci. 40*: 327 (1975).
52. R. Miller and H. S. Groninger, Jr., *J. Food Sci. 41*: 268 (1976).
53. K. L. Franzen and J. E. Kinsella, *Agric. Food Chem. 24*: 788 (1976).
54. E. A. Childs and K. K. Park, *J. Food Sci. 41*: 713 (1976).
55. D. R. Grant, *Cereal Chem. 50*: 417 (1973).
56. S. K. Gandhi, J. R. Schultz, F. W. Boughey, and R. H. Forsythe, *J. Food Sci. 33*: 163 (1968).
57. A. J. King, H. R. Ball, and J. D. Garlich, *J. Food Sci. 46*: 1107 (1981).
58. Y. Sato and R. Nakamura, *Agric. Biol. Chem. 41*: 2163 (1977).
59. C.-Y. Ma and J. Holme, *J. Food Sci. 47*: 1454 (1982).

60. C.-Y. Ma, L. M. Poste, and J. Holme, *Can. Inst. Food Sci. Technol. J. 19*: 17 (1986).
61. J. G. Montejano, D. D. Hamann, H. R. Ball, Jr., and T. C. Lanier, *J. Food Sci. 49*: 1249 (1984).
62. H. N. Singh and W. L. Hinze, *Analyst 107*: 1073 (1982).
63. F. Arranz and M. Sander, Chaves, *Rev. Plast. Mod. 42*(305):535 (1981).
64. A. A. Ansari, S. A. Kidwai, and A. Salahuddin, *J. Biological Chem. 28*: 1925 (1975).
65. V. P. Torchilin, V. G. Omel'yanenko, A. L. Klibanov, A. I. Mikhailov, V. I. Gold'danskii, and V. N. Smirnov, *Biochim. Biophys. Acta 602*: 511 (1980).
66. A. Anthony, M. Schafidi, and M. Blatrir, *Cosmetics and Toiletries 95*: 65 (1980).
67. Z. Haque, T. Matoba, and M. Kito, *J. Food Chem. 30*: 481 (1982).
68. S. Magdassi, D. Leibler, and S. Braun, *Langmuir 6*(2): 376 (1990).
69. S. Magdassi and A. Stavsky, *J. Dispersion Sci. Tech. 10*(3): 713 (1989).
70. S. Magdassi, A. Stavsky, and S. Braun, *Tenside Surf. Det. 28*: 164 (1991).
71. S. M. Sullivan and L. Huang, *Biochim. Biophys. Acta 812*: 116 (1985).
72. A. Huang, Y. S. Tsao, S. J. Kennel, and L. Huang, *Biochim. Biophys. Acta 716*: 100 (1962).
73. Z. Haque and M. Kito, *Agric. Biol. Chem. 46*(2): 597 (1982).
74. S. Arai and M. Watanabe, in *Advances in Food Emulsions and Foams* (E. Dickenson and G. Stainsby, eds.), Elsevier Applied Science, 1988, p. 189.
75. C. A. Kuehler and C. M. Stine, *J. Food Sci. 37*: 604 (1972).
76. M. Yamashita, S. Arai, Y. Imaizumi, Y. Amano, and M. Fajimoki, *J. Agric. Food Chem. 27*(1): 52 (1979).
77. M. Watanabe, H. Toyokawa, A. Shinoda, and S. Arai, *J. Food Sci. 46*: 1467 (1981).
78. Z. Haque and M. Kito, *J. Agric. Food Chem. 32*: 1392 (1984).
79. M. T. A. Evans, J. Mitchell, P. R. Mussellwhite, and L. A. Irons, *Adv. Exp. Medicine Biology 7*: 1 (1970).
80. A. Shimada, I. Yamamoto, H. Sase, Y. Yanaraki, M. Watanabe, and S. Arai, *Agric. Biol. Chem. 48*(11): 2681 (1984).
81. Z. Haque and M. Kito, *J. Agric. Food Chem. 31*: 1231 (1983).
82. M. Gresen-Wiese, Proceedings of Incosmetic Conference, Paris, 1995.

83. S. Magdassi, O. Sheinberg, and Z. Zakay-Rones, ACS Symposium Series, *Proteins at Interfaces,* (J. Brash, ed.), 1995.
84. A. V. Kabanov, E. V. Batrakova, N. S. Melik-Nubarov, N. A. Fedoseev, T. Y. Dorodnich, V. Y. Alakhov, V. P. Chekhonin, I. R. Nazarova, and V. A. Kabanov, *J. Controlled Release 22*: 141 (1992).
85. S. Magdassi, O. Toledano, and Z. Zakay-Rones, *J. Colloid Interface Sci., 175*: 435 (1995).
86. S. Magdassi, Z. Zakay-Rones, and O. Toledano, PCT/EP94/02827, 1994.

3
Increased Anionic Charge
Conformational and Functional Properties

K. D. Schwenke

University of Potsdam, Bergholz-Rehbrücke, Germany

I. INTRODUCTION

Already in 1949 proteins were chemically modified by reaction with acetic anhydride for biochemical studies [1]. By analogy with this reaction, succinic anhydride was introduced [2]. By attaching the free amino groups of proteins, this reagent converts a positively charged group into a negatively charged succinyl function. It was pointed out in 1958 that, owing to the increased anionic charge, succinylation was more effective in inducing conformational changes in proteins [3]. With maleylation [4,5] and citraconylation [6], possibilities of reversible blocking of protein amino groups were introduced in 1967–1969.

Studies on the succinylation of chicken myosin in 1966 and 1967 [7,8] opened the investigation of chemically modified food proteins bearing increased anionic charge. The chemical modification of egg white with 3,3-dimethylglutaric anhydride in 1968 [9] was one of the first examples where protein acylation was performed to alter the functionality of a food product. The improvement of functional properties was the aim of food protein acylation both by acetylation and by succinylation, which was carried out in the following years on a large number of proteins of animal and vegetable origin [10–20]. In most cases, the change of emulsifying and foaming properties, in-

61

duced by the modification, has been studied as an important indicator of improved functionality. The modified surface functionality became especially interesting for attempts to correlate functional and molecular structural properties of proteins [21,22]. Since proteins can be modified in different ways by different amounts of acylating agents, the structure of a protein may be modified step by step. This can be very helpful in finding structure–functionality relationships. Moreover, this kind of modification should be useful in defining protein functional properties on a scientific basis. On the other hand, it allows us to create "tailored" protein products for a large spectrum of applications. While the acylation of proteins was intended at the beginning of this development to provide products for food and feed purposes, it is now more and more focussed on the field of nonfood applications.

II. ATTACHMENT OF ACYL RESIDUES TO PROTEINS

A. Chemical Backround

The different reagents employed to acylate proteins do not react selectively with one type of functional group but can react with all nucleophilic groups. These comprise amino groups (N-terminal α- and lysine ε-amino groups), phenolic (tyrosine) and aliphatic (serine and threonine) hydroxyl groups, sulfhydryl groups (cysteine) and imidazole (histidine) groups. However, both the reactivity of these groups and the stability of their acyl derivatives differ considerably.

The ε-amino group of lysine is the most readily acylated group because of its relatively low pK and its steric availability for reaction. Since the protonated and the unprotonated form of a reactive side chain have largely different chemical properties, pH has an important influence upon the acylation of this group. Generally, reactivity increases with increasing nucleophilicity and decreasing protonation. The rate of acylation is therefore highly reduced or even suppressed when the pH is significantly lower than the pKa of the functional group. While imidazole groups are more reactive in neutral solution than amino groups, due to the protonation of the latter, amino groups become the most reactive ones at higher pH's. However, the acylating reagent, carbonic acid anhydride, is quickly hydrolyzed in alkaline solution, resulting in a drop in the extent of acylation of amino groups. Therefore the acylation of amino groups is practically performed under weakly alkaline conditions (pH 7.5–8.5).

Succinylation of amino groups results in the formation of N-succinyl (Fig. 1(1)) bonds, which are very stable in neutral as well as weakly acidic and alkaline conditions. In contrast to this, O-succinyltyrosine residues, which are readily formed under the conditions of amino group succinylation, are not stable, owing to a rapid intramolecular hydrolysis above pH 5 (Fig. 1(2)). Serine and threonine hydroxyl groups are not easily acylated in aqueous solutions, because they are weak nucleophiles. However, their succinyl derivatives are rather stable and do not hydrolyze as spontaneously as O-succinyltyrosine, but they can be cleaved by treatment with hydroxylamine [24]. Acylation of histidine and cysteine residues yields products that are highly unstable and are readily hydrolyzed in aqueous solution [23].

Maleic anhydride, 2-methylmaleic anhydride (citraconic anhydride), and 2,3-dimethylmaleic anhydride react with proteins in a manner similar to succinic anhydride, but the products are much more labile to hydrolysis [5,6]. The reaction products of maleic anhydride and amino groups are stable at neutral pH but rapidly hydrolyze when acidified to pH 3.5. This lability increases in the order maleyl < citraconyl < dimethylmaleyl derivatives. Citraconylamides are stable enough to withstand many hours at neutral or alkaline pH's but

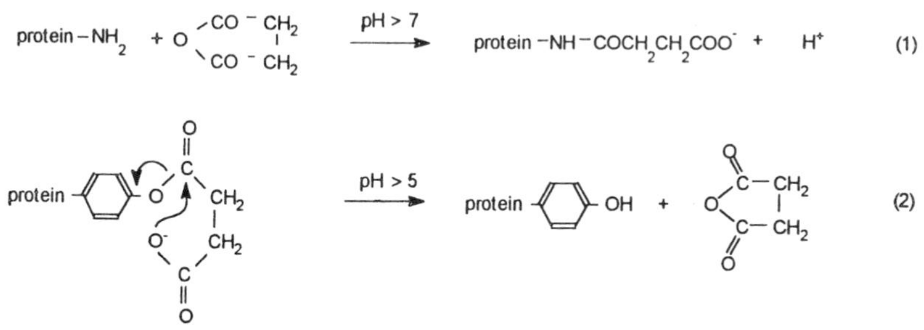

Figure 1 Succinylation of amino groups (1) and desuccinylation of O-succinyltyrosine (2) in proteins.

are sufficiently labile at low pH's to be deacylated under mild conditions [6].

B. Extent of Blocking of Functional Groups

Amongst the different protein derivatives bearing an increased anionic charge, highly stable *N*- and *O*-succinyl proteins are most important for potential practical use. Since the physicochemical properties of the modified proteins depend essentially on the amount of introduced residues, the extent of blocking of each type of functional group is a characteristic parameter of acylated proteins. Steric and proximity effects of topologically adjacent residues frequently prevent individual residues from reaction. Therefore acylation of native proteins seldom leads to the complete blocking of amino groups.

Using a 12-fold molar excess of succinic anhydride, 90–94% of the amino groups of three globular proteins, bovine γ-globulin, bovine serum albumin, and β-lactoglobulin, were blocked [3]. A 90% blocking of amino groups was reached by succinylation of chicken heart myosin [8], while only 77% of the amino groups of myofibrillar fish proteins could be succinylated [10]. About 87% of the amino groups of beef heart myofibrillar proteins could be blocked with succinic anhydride and about 98% with 1,2,4-benzenetricarboxylic anhydride [25]. The oligomeric 11 S storage proteins from plant seeds, though very similar in structure [26], show a rather different reactivity of amino groups with succinic anhydride. While a maximum succinylation of 82% and 85% could be attained with arachin from peanuts [27] and helianthinin from sunflower seeds [28], more than 95% of the amino groups of pea legumin was succinylated [29]. The lowest level of acylation was achieved at a comparable excess of reagent with cruciferin from rapeseed amounting to 70% (30). A level of 80–90% succinylation of cruciferin could be obtained only when the excess of reagent was increased to more than 200-fold (Fig. 2). It will be shown in Sec. III that this increase in acylation is due to conformational changes of the protein, which result in a rather unfolded structure and in an increasing availability of reactive residues [31].

As can be seen in Fig. 2, a high amount of amino groups in pea legumin is very reactive and is succinylated at only a small excess of reagent. Only traces of succinylated OH groups were determined un-

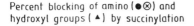

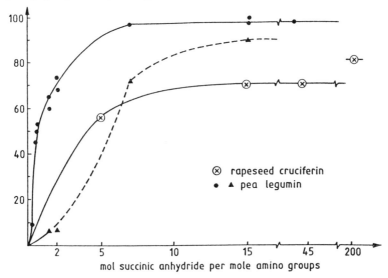

Figure 2 Succinylation of amino and hydroxyl groups in pea legumin and of amino groups in rapeseed globulin (cruciferin) as a function of the amount of succinic anhydride. (From Refs. 29 and 30.)

der these conditions. The blocking of OH groups (esterification) became, however, significant at a high level of *N*-succinylation (about 75%) and reached 100% at a 20-fold molar excess of succinic anhydride over the total amount of amino and hydroxyl groups [29]. This finding is consistent with the results obtained on succinylated pepsinogen [24], where the acylation of serine and threonine in proteins was observed first. The authors concluded that "modification of the hydroxy amino acids is at a minimum when unreacted lysine and tyrosine residues are still present, but it increases quite sharply when the reaction with the lysyls and tyrosyls is essentially complete."

The succinylation of 20% of the total hydroxy amino acids in bovine serum albumin occurred under conditions under which most of the lysyl residues were succinylated [32]. Although succinyltyrosine undergoes a hydrolytic splitting at pH > 5, the blocking of tyrosine residues was found to be significant in the case of succinylcarboxypeptidase [33] and succinylpepsinogen [24]. A complete recovery of tyrosyl hydroxyl groups was achieved in these cases after some hours in neutral solution.

Although S-acyl derivatives are thought to be unstable, the blocking of cysteinyl residues during the succinylation of some proteins was reported. Using an 800-fold molar excess of succinic anhydride, a 40% blocking of sulfhydryl groups of actin was found, whereas only 30% of the ϵ-amino groups of lysine were succinylated and the tyrosyl groups remained unchanged [34].

The acylation of functional groups other than amino groups was scarcely investigated in food proteins. In a comparative study with beef heart myofibrillar proteins using different acylating reagents, a decreasing reactivity with OH groups in the order acetic anhydride > succinic anhydride > cis,cis,cis, cis-tetrahydrofuran-2,3,4,5-tetracarboxylic dianhydride > 1,2,4-benzenetricarboxylic anhydride was found [25]. The high blocking of SH groups with acetic anhydride (80%) and the low extent of S-succinylation (25%) are a hint to the quicker hydrolysis of the latter.

The study of succinylation of cheese whey protein concentrates gave a blocking of 83% of the amino groups, 51% of sulfhydryl, and 27% of hydroxyl groups [35].

III. CONFORMATIONAL CHANGES

A. Small and Medium-Size Globular Proteins Consisting of One Polypeptide Chain

The introduction of succinyl residues producing short-range repulsive forces in place of possible short-range attractive forces in the native molecule resulted in a change of hydrodynamic properties of bovine serum albumin (BSA), bovine γ-globulin, and β-lactoglobulin [3]. The succinylated derivatives showed markedly increased intrinsic viscosity and Stokes radius and a decrease of sedimentation coefficient [36,37]. These results are compatible only with a considerable increase in the effective volume occupied by the succinylated protein molecule compared to its unreacted counterpart.

Further evidence for the rather unfolded state of succinylated proteins is provided by spectroscopic [24,37] and chemical studies [37].

The succinylated proteins showed an enhanced susceptibility to sulfitolysis owing to the increased exposition of buried disulfide groups
[37]. Using circular dichroism (CD) spectroscopy a decrease of helical content in the secondary structure of BSA was observed [32]. The
view that the macromolecular conformation of succinylated proteins
is loosened and resembles that of partially unfolded proteins was also
supported by the investigation of succinylated pepsinogen using optical rotatory dispersion [24]. The study of the course of succinylation
of BSA and lysozyme in dependence on the amount of succinic anhydride revealed two different ranges of succinylation with regard to
conformational changes [37]. Small changes of the Stokes radius,
disulfide susceptibility, and capacity of antibody precipitation were
observed at moderate degrees of succinylation. A sudden increase of
Stokes radius and disulfide sulfitolysis as well as a decrease in antibody precipitation occurred at a higher degree of succinylation, indicating the beginning of unfolding [37]. Due to structural peculiarities,
this "degree of unfolding," corresponding to a certain amount of introduced negative charges, differs between BSA and lysozyme. The
comparison of the sedimentation coefficients of BSA unfolded in 6
M guanidinium chloride and of exhaustively succinylated BSA led to
the conclusion that the unfolding due to succinylation was still
limited.

Using fluorescence polarization, rotational relaxation time, and
sedimentation velocity measurements it could be shown that BSA under certain conditions of succinylation, maleylation, or citraconylation exists in an expanded form, very similar in its physical properties
to bovine serum albumin expanded by electrostatic repulsion at pH 2
[38]. The expansion was completely reversed at a sufficiently high
ionic strength, which compensated the electrostatic effect of introduced negative charge. The dependence of both the sedimentation
coefficient and the fluorescence polarization on the extent of N-acyla-
tion revealed a critical degree of modification (about 40%) above
which conformational changes appeared [38].

B. Myosin

Succinylation of myosin resulted in a rise of intrinsic viscosity at low
degrees of modification, which indicated expansion or swelling of the

molecule or both. On further succinylation it fell to the level of the nonsuccinylated control myosin; thus the molecule seemed to assume a more compact shape [8]. Light scattering data revealed a slight increase of the molecular weight of myosin after extensive succinylation from 5.87×10^5 to 6.20×10^5 Da, which was obviously due to the amount of attached succinyl residues. The molecular dimensions of the modified protein were comparable to these of myosin. Measurements of optical rotation indicated, however, a continuous decrease of the helical content of myosin and transformation into a more random conformation. Ultracentrifugation patterns revealed the presence of two components, the main component of which had hydrodynamic properties very close to those of nonsuccinylated myosin. The minor component had a molecular weight of 4.2×10^4 Da and was said to be dissociated from the globular head of heavy meromyosin [8]. Full succinylated myosin was water-soluble and remained so even after prolonged heating. It completely lost both its Ca^{2+} activated ATPase activity and its ability to combine with F-actin, already at low degrees of succinylation.

C. Oligomeric Enzymes and Related Proteins

The net electrostatic repulsion induced by extensive succinylation of proteins resulted in a dissociation of most oligomeric proteins into subunits [39–45]. Succinylation as a mild procedure is superior to some other methods of protein dissociation. It was, therefore, successfully applied for the determination of the subunit molecular weight and thereby for elucidating the subunit arrangement in the quaternary structure of proteins such as hemerythrin, the oxygen-carrying protein of marine invertebrates [39], aldolase [40], and aspartate-transaminase [41]. Maleylation induced formation of subunits of molecular weight half of that of the original molecule of the 81 kDa ATP creatine phosphotransferase [46].

There are some oligomeric proteins—for instance α-crystallin [47] and high-density lipoprotein [48]—which did not dissociate into subunits, though 90% of lysyl residues were succinylated. Succinylation of aldolase [40] and ceruloplasmin [43] in neutral solution led to intermediary subunits only, whereas the final dissociation into the smallest subunits did not occur until the negative charge was further increased by enhancing the pH. Similarly, the dimeric enzyme aspartate-transaminase after being succinylated underwent a complete dissociation into two subunits only in alkaline solution [41].

Succinylation of oligomeric proteins is a valuable tool for hybridization studies, which are particularly useful for investigating association–dissociation equilibria in proteins and for elucidating relationships among primary, secondary, tertiary, and quaternary levels of structure. For hybridization studies, only those weakly succinylated proteins are appropriate that are homogenous and still posses the quaternary structure analogous to that of the native enzyme. Hybridization consists in mixing an unmodified protein with its succinylated counterpart in a dissociated and unfolded state. During renaturation hybridized molecules are formed consisting of an unmodified and a succinylated part in a different composition. The electrophoretic mobility of the hybridized form is higher than that of the native and lower than that of the succinylated molecule. Since the number of electrophoretic zones depends on the number of subunits, the subunit composition of an enzyme can be determined. This was carried out successfully on some proteins such as aldolase [49], aspartate transcarbamylase [50], and hemerythrin [51].

D. Oligomeric Storage Proteins from Plant Seeds

The main storage proteins from dicotyledonous plants are represented by two types of oligomeric proteins, the legumin-like 11-S-globulins and the vicilin-like 7-S-globulins [52]. The 11-S-globulins with molecular masses between 300 and 360 kDa are composed of six structural nearly identical subunits, each consisting of two unequal disulfide-bridged polypeptide chains, an acidic 30–40 kDa α-chain and a basic 20 kDa β-chain [52]. They are homologeous proteins and form a very similar quaternary structure [26,53]. The 7-S-globulins have a trimeric structure [54]. The combination of 50 kDa and 70 kDa subunits can lead to molecular masses between 150 and 210 kDa for the oligomeric 7-S-protein [55]. Succesive succinylation of 11-S-globulins induces a stepwise decay of the oligomeric molecule into the 3–

4 S subunits. This dissociation has thoroughly been studied in the case of the corresponding proteins from peanuts ("arachin") [27], sunflower seeds ("helianthinin") [28], rapeseeds ("cruciferin") [30], and peas ("legumin") [29]. Ultracentrifugation studies revealed a general course of the dissociation corresponding to the pathway

11–13 S   → 7–9 S   → 3–4 S
hexamer   trimer   monomer
300–360 kDa   150–180 kDa   50–60 kDa

This is shown in Fig. 3 for the dissociation of helianthinin [28]. In the case of arachin [27] and cruciferin [21], small amounts of a 3–4 S component were found besides the 7–9 S intermediate fraction already at an early stage of dissociation of the weakly succinylated proteins. In these cases it is not clear whether the monomeric dissociation products (3 S component of cruciferin, 4 S component of arachin) derive from the 7 S or 9 S half molecule or directly from the original protein. These data point to differences in the stability of the subunit structure amongst the 11-S-globulin. Moreover, they may be a hint to a favored reactivity of some protein subunits.

The investigation of the succinylated pea legumin revealed that the weakly acylated protein did not dissociate and only 20% of the protein was found to be dissociated as 7 S half molecule, when about 45% of the lysine residues were succinylated [29]. These correspond approximately to the number of lysines exposed at the molecular surface [31]. The disappearance of the 12 S and 7 S components of helianthinin in favor of a 3 S component (Fig. 3) coincides with a sudden increase of the intrinsic viscosity and a change in the near ultraviolet CD spectra. Analogous changes were found at a distinct level of succinylation for the other studied 11-S-globulins, cruciferin (Fig. 4) [30], arachin [27], and pea legumin [29]. These changes reflect a cooperative process leading to a sudden unfolding of the dissociated subunits at a certain critical level of succinylation. The latter corresponds to 60–70% succinylation of the amino groups of arachin, cruciferin, and helianthinin and to about 70–80% for pea legumin. It is an indication of a different stability of subunit conformations and interactions between both polypeptide chains in the various proteins.

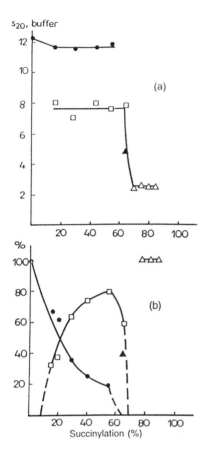

Figure 3 Ultracentrifugal investigation of the dissociation of helianthinin from sunflower seeds in dependence on the degree of N-succinylation. (a) Sedimentation coefficients. (b) Percentage of the components of dissociation. ● 12 S. □ 7 S. ▲ 5 S. △ 2.5 S. (From Ref. 28.)

The unfolding of polypeptide chains is favored by the increase of negatively charged groups due to the succinylation of hydroxyl groups, which became significant after reaching the critical level of N-succinylation (Fig. 2) [29]. Succinylation makes these proteins more hydrophilic. The hydrophobicity of pea legumin measured by fluorescence probe techniques decreased with the increasing number of introduced succinyl residues [56].

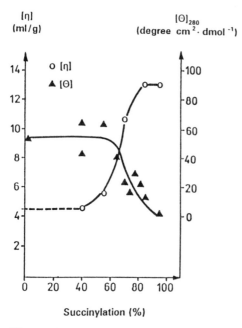

Figure 4   Dependence of the intrinsic viscosity [η] and the mean residue ellipticity at 280 nm [θ] of cruciferin from rapeseed on the degree of N-succinylation. (From Ref. 30.)

Evidence for the alteration of the tertiary structure of arachin and pea legumin at moderate degrees of succinylation was provided by fluorescence spectra [27,29]. Both proteins showed an increase of fluorescence intensity up to 60% N-succinylation (Fig. 5) pointing to the formation of a new structure in which the chromophoric residues (essentially tryptophan) are buried and the quantum yields are higher. The strong decrease in the fluorescence intensity at a step of 75% N-succinylation of legumin and the red shift of the fluorescence maximum are arguments for the unfolding process also documented by the increase of intrinsic viscosity at the same level of modification [29].

The investigation of maleylated vicilin, the 7-S-globulin from faba beans, revealed the existence of a critical degree of modification for the unfolding of this type of protein, too [57]. While no significant changes in denaturation temperature and enthalpy of denaturation were found up to 63% modification, these values decrease signifi-

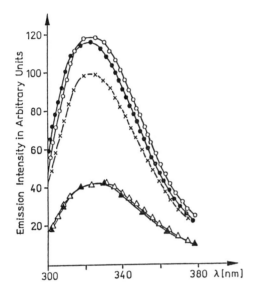

Figure 5   Fluorescence emission spectra of pea legumin at various steps of N-succinylation; x, unmodified; ○, 50%; ●, 60%; △, 75%, and ▲, 95% succinylation. (From Ref. 29.)

cantly at higher degrees of maleylation, reaching the lowest values at 94% modification. The hydrophobicity of vicilin decreased with increasing level of maleylation up to the critical point (63%) and increased continously with the extent of modification at higher levels of acylation. In this case, the intrinsic effect of newly exposed hydrophobic regions exceeds that of introduced hydrophilic maleyl residues. This result is in contrast to the data on succinylated legumin and point to structural peculiarities of the proteins and to the difficulty of predicting the hydrophobicity on the basis of the amount of introduced acyl residues only.

The existence of a critical point of modification indicating sudden changes of the tertiary structure both in bovine serum albumin [37,38] and in oligomeric plant proteins points to a rather similar behavior of globular proteins in succinylation or related acylation.

Secondary structure changes in succinylated arachin, helianthinin, and legumin have been studied using far-UV CD spectroscopy. In the cases of arachin [27] and helianthinin [31], a significant increase of the negative ellipticity values around 210 nm occurred after passing the critical degree of modification (>65%). Though such a spectral change would normally be interpreted as an increase of helical conformation, a corresponding explanation in case of highly negatively charged polypeptide chains seems to be difficult. Since aromatic side chains can contribute to CD spectra not only in the near-UV but also in the far-UV region [99,100], these side chain contributions should be taken into account for understanding the far-UV CD spectra of succinylated arachin and helianthinin. Accordingly, the far-UV CD spectra of both proteins would reflect the changes of the secondary and tertiary structure. Aromatic side chain contributions to far-UV CD spectra have been especially observed for the molten globule state of proteins [101].

In contrast to arachin and helianthinin, CD spectroscopic studies of succinylated legumin revealed a decrease of helical conformation in highly modified samples [29]. These findings were confirmed by infrared spectroscopic investigations [58]. The different spectroscopic properties of succinylated helianthinin and arachin on the one hand and succinylated legumin on the other hand point to structural differences of these proteins.

IV. SURFACE FUNCTIONAL PROPERTIES

A. Emulsifying Properties

Introduction of additional negative charge into proteins by reaction with dicarboxylic anhydrides resulted in an increased solubility in neutral and weak alkaline aqueous solutions [11–13]. A positive correlation between solubility and the ability of a protein to emulsify has been documented by some authors [59–62]. An increase in protein solubility would encourage a rapid migration to and adsorption of the protein at the water–oil interface. The adsorption would, in turn, lower the interfacial tension between the water and the oil and stabilize the emulsion [63]. In fact, improved emulsifying properties have been found after succinylation or citraconylation of a large number of food proteins such as myofibrillar fish proteins [10], fish protein concentrate [11], beef heart myofibrillar proteins [25], egg white [64], casein [65], wheat gluten [19], protein isolates or concentrates from oat [66], peanut flour [13], cottonseed [17,67–69], soybean [12,15], sunflower seed [16,70,71], field pea [72], faba bean [18], winged bean [73], rapeseed [74–77], and green leaves [78]. The increase of emulsification parameters (emulsifying capacity (EC), emulsifying activity (EA), or emulsifying activity index (EAI) and emulsion stability (ES)) was not generally linearly related to the increase in the extent of succinylation (for instance) or to the increase in nitrogen solubility. This is because the emulsifying property is dependent not only on the solubility but also on the hydrophile–lipophile balance (HLB) of the particular protein. If the HLB of the protein is close to the optimal HLB for the oil, EC, EA, and ES should be high. However, it is rather difficult to define the term HLB for a protein. The term hydrophobicity has been introduced and experimentally studied for proteins in order to find a relationship between structural and functional properties of proteins [79,80].

Multiple regression analysis performed for succinylated rapeseed protein isolates indicated that emulsification activity was related to protein solubility, hydrophobicity, zeta potential, and flow behavior of aqueous dispersions of the proteins. Emulsion stability was affected by protein solubility, zeta potential, apparent viscosity of protein dispersions, and difference in density between aqueous and oil phases [76].

Structural peculiarities of the different proteins should be therefore taken into account when a relationship between the degree of succinylation and emulsification is derived. These include conformational changes and the contribution of hydrophobic regions exposed by the unfolding of the globular protein structure. Moreover, the technological treatment of the raw material or even the protein, which can induce denaturation or interaction with nonprotein compounds, is to be considered for understanding the possibly different behavior of proteins from the same source.

pH and ionic strength are most important extrinsic factors influencing the charge of a protein and the efficiency of acylation in affecting functionality. The shielding effect of high salt concentration can counteract even the expansion effect of high succinylation in proteins [38]. In view of the importance of a neutral or weak acidic milieu for food systems, most functional studies were undertaken under corresponding conditions of pH.

High or even exhaustive modification has been reported to be most effective for a number of proteins in increasing emulsification parameters at a pH higher than the I.P. This was shown for proteins from eggs [64], myofibrillar proteins from fish [10] and beef heart [25], proteins from leaves [78], soybeans [12,15], peas [72], faba beans [18], oats [66], and others. Protein isolates from faba beans consisting of 73% legumin and 27% vicilin showed a continuous increase of emulsification parameters in neutral solution with increasing degree of succinylation [18] (Fig. 6). Succinylation of a soybean protein isolate over 95% improved the ES at pH 7 by a factor of 2.5 compared to that of untreated proteins. Succinylation between 30 and 70% increased the ES at pH 3.5 by 40%, but succinylation over 90% decreased the ES at this pH drastically [15]. A significant increase in EC of peanut flour was noted to occur at medium level of succinylation, while the highest level of modification resulted in a significant reduction in EC [13]. Protein isolates and concentrates from sunflower and rapeseed consisting mainly of the corresponding 11-S-

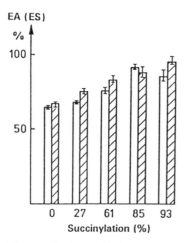

Figure 6   Emulsifying activity (EA, □) and emulsion stability (ES, ■) of native and succinylated faba bean protein isolates in dependence on the degree of N-succinylation. (From Ref. 18.)

globulins showed the highest increase of EC, EA, and ES at low or medium levels of succinylation [16,70,71,74,77]. The EA of succinylated rapeseed protein isolate increased with the enhancing of the salt concentration up to 0.35 M NaCl. The ES, however, decreased with increasing ionic strength [76]. Investigation of the emulsification of the purified rapeseed 11-S-globulin, cruciferin, at pH 9.2 and medium ionic strength (I = 0.15), revealed a significant emulsion stabilizing effect of succinylation. It increased with the degree of modification, reaching the maximum value at exhaustive blocking of amino groups [21]. Napin, a low-molecular-weight basic storage protein from rapeseed [81,82], is very hydrophilic and did not give an improvement of emulsifying properties after introduction of an excess of anionic groups. Succinylation lowered the EA and ES of napin [83] and reduced its capacity to decrease interfacial tension [84].

Succinylation of casein reduced the EA and ES but increased the EC significantly at pH 7 [65]. The latter showed a continuous growth with the extent of modification at 37–97% succinylation. Measurements at pH 4, where the solubility of casein and succinylated casein is minimal, revealed extremely low values of EC [65].

B. Foaming Properties

Proteins with a high foam forming capacity are able to reduce the surface and interface tension of aqueous liquids and to form structurally continuous, cohesive films around air vacuoles [85]. Like emulsifying properties, foaming properties depend on the solubility, charge, and hydrophobicity of the protein [79,80]. Moreover, they are significantly influenced by the viscosity of the protein solution [79,80]. Succinylation that alters all these parameters has a positive effect upon the foam expansion (foam capacity = FC) of a number of food proteins. The FC was enhanced by succinylation of protein preparations from leaves [78], soybean [12], sunflower [16,70,71], faba bean [18], pea [72], cottonseed [17,68,69], and oat [66] proteins and others. Some proteins that have high FC in unmodified state did not alter the FC or showed a reduced FC after succinylation. Napin, the low-molecular-weight storage protein from rapeseed, for instance, gave extremely high FC values in weak acidic and neutral solutions.

Succinylation from 17% to 90% did not change the FC [83,84]. Egg white, the "classical" foam forming protein, showed a drop of FC after succinylation [64]. Succinylation reduced, however, heat-induced damage of the foaming properties of egg white [14].

Inconsistent data on foaming properties of rapeseed protein concentrates and isolates [74,75] underline the important influence of the technological process used to obtain the protein preparations upon FC. Thus native rapeseed globulin, which gave a very high FC comparable to that of napin, did not show a reduction of FC after moderate or high succinylation [75]. Succinylation of rapeseed meal prior to protein extraction resulted, however, in modified isolates with poor foaming properties [74].

The contribution of viscosity to foam formation has been shown with succinylated faba bean protein isolates both by increasing the protein concentration and by inducing conformational changes by heating or succinylation [86,87]. The FC of these proteins reached maximal values in 3 or 5% solutions only at exhaustive succinylation. Augmenting the protein concentration to 10% and additional heating increased the EC to maximal values also at low and moderate succinylation. However, when the degree of modification was high enough to cause protein unfolding and viscosity increase [87], maximal FC was attained even with unheated solutions [86] (Fig. 7).

While foaming is favored by increased viscosity, hydrophobicity, and solubility, the increase of net charge density caused by succinylation tends to decrease foam stability (FS), since it prevents optimum protein–protein interactions required in a continuous film around air bubbles. Therefore a number of succinylated proteins showed a drop of the FS with increasing degree of modification. This was reported for protein preparations from faba beans [86], peas [72], cottonseed [69], oat [66], and cheese whey protein concentrate [88] for instance. The unfavorable charge effect of succinylation can be overcome by foaming at sufficiently high protein concentration, as shown for faba bean protein isolates [86] and napin from rapeseed [84].

In some protein preparations, such as sunflower isolate and concentrate [16,70] and soybean protein [12], stable foams were obtained, although the extent of succinylation was high. In these cases, the other intrinsic and extrinsic parameters obviously overcome the charge repulsion. The investigation of foam destabilization kinetics

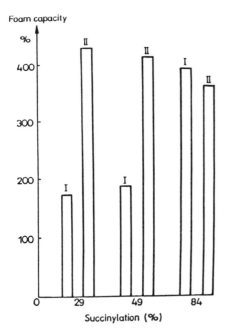

Figure 7 Foam capacity of succinylated faba bean protein isolates at different degrees of modification. I, without heating. II, after heating to 80°C prior to foaming. (From Ref. 86.)

of variously succinylated rapeseed globulin, cruciferin, revealed the lowest destabilizing effect at the highest degree of succinylation (83%) [21]. The corresponding foam stability index (FSI) was discussed as a measure of coalescence phenomena. Accordingly, succinylation mainly improved the efficiency of the protein in limiting the coalescence by electrostatic repulsion between air bubbles.

C. Attempts to Correlate Conformational Changes and Surface Behavior of the Succinylated 11-S-Globulins Cruciferin and Legumin

Taking into account the considerable difficulty in correlating structural and functional parameters of proteins and—still more—in quantifying such relationships, an attempt was made to study the kinetics

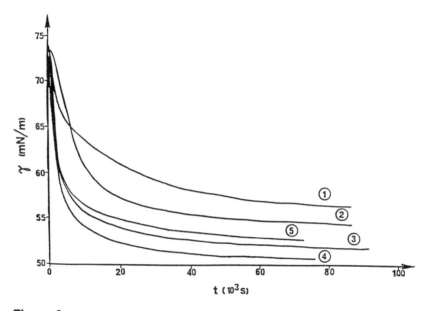

Figure 8 Adsorption kinetics (interface air–water) of cruciferin from rapeseed at various degrees of succinylation (pH 9.2, 1 = 0.15). 1, unmodified; 2, 19%; 3, 60%; 4, 66%; and 5, 84% succinylated. γ = interfacial tension. (From Ref. 21.)

of surface adsorption of rapeseed cruciferin [21] and pea legumin [22] at various degrees of succinylation, i.e., at different states of conformation. The kinetics of adsorption at the air–water interface showed a decrease of interfacial tension at the equilibrium (corresponding to a rise of interfacial pressure Π_e) with increasing degree of modification up to 66% and 75% succinylation for cruciferin [21] and legumin [22] respectively (Fig. 8). Further enhancement of the extent of succinylation led to the decrease of Π_e. Since this inversion of the interface behavior occurred just after passing the critical points of succinylation (*vide* Sec. III.D), it is likely to be related to the unfolding of the proteins and to the drastic increase of negative charge by additional O-succinylation [29,58] (Fig. 9).

To get more information about the adsorption process, the three-step concept of kinetics [89–91], assuming a diffusion step to the interface, a step of penetration through the interface film, and a step of molecular rearrangement within the adsorbed layer, was applied [21,22]. The commonly used plots of Π and Π_e (interfacial pressure at time t and at equilibrium) versus time t, which are $\Pi = f(t^{1/2})$, In $(d\Pi/dt) = f(\Pi)$, $\ln(1 - \Pi/\Pi_e) = f(t)$ [92–95], were applied for the analysis of the different kinetic steps. The plot of Π vs. $t^{1/2}$ showed a linear region, giving the limits of the diffusion controlled step [93]. The kinetic rate constant of diffusion of variously succinylated cruciferin samples increased from 1.9×10^{-4} s^{-1} for the native protein to 5.0×10^{-4} s^{-1} for the 83% succinylated sample [21]. This result reflects the decreased molecular size of cruciferin due to increasing dissociation of the protein. A more detailed description of the dependence of kinetic constants on the extent of succinylation was given for the three kinetic steps with succinylated legumin [22] (Fig. 10).

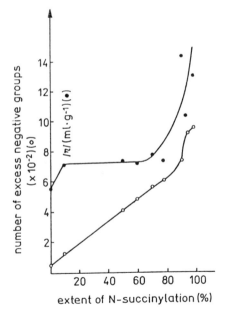

Figure 9 Dependence of the intrinsic viscosity (η) (●) and the number of excess negative groups (excess protein carboxyl groups and introduced succinyl residues) (○) of pea legumin on the degree of N-succinylation. (From Ref. 58.)

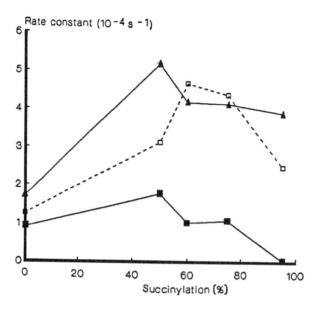

Figure 10   Variation of rate constants for the diffusion step k1 (▲), penetration step k2 (□), and rearrangement step k3 (■) of the adsorption of
native and succinylated pea legumin at the air–water interface in dependence
on the degree of N-succinylation. (From Ref. 22.)

The rate constants of the diffusion (k1) and penetration (k2) steps
are significantly higher for the modified samples than for the native
protein. The highest values of k1 were obtained for 50% succinylated
samples consisting of about 20% dissociated 7 S and 80% undissoci-
ated 12 S component [29]. However, as revealed by fluorescence
(Fig. 5) and CD (Fig. 11) spectroscopic investigations, the legumin
underwent discrete changes of tertiary structure already at this step of
modification. One can assume that the increased adsorption rate of
these samples was due to a raised accessibility and flexibility of external loops, which became more active in the anchorage process of
the protein at the interface. Consequently, it can be concluded that
both diffusion and ability to anchor must be considered in step 1. At
higher degrees of succinylation, dissociation into subunits became
more important. Most interesting is, however, the decrease of k2
after passing the critical step of unfolding at 75% succinylation. This

reflects the hindered penetration of the highly charged and unfolded molecules due to electrostatic repulsion and steric effects. The latter may be responsible also for the slowdown of the speed constant of the rearrangement step (k3) at exhaustive succinylation, although the changes of k3 were less significant.

Charge and conformation effects became obvious also, when another kinetic parameter, the energy barrier ΔA, was analyzed. ΔA was determined by Tornberg [93] as the mean area necessary for a molecule to adsorb. It is noteworthy that a significant increase of ΔA occurred for the penetration step (A2) also after passing the critical degree of succinylation (Fig. 12). High charge and molecular expansion caused a high-energy barrier against the penetration through the interfacial layer. Conformational effects should be considered for the interpretation of change of A3 (rearrangement). In contrast to both step 2 and step 3, the energy barrier of the diffusion step 1 is negligible.

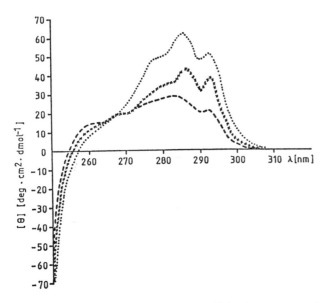

Figure 11   Near-ultraviolet circular dichroism spectra of unmodified (....), 50% succinylated (xxx), and 95% succinylated (---) pea legumin. (From Ref. 29.)

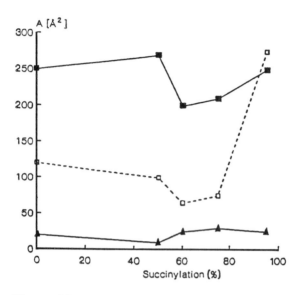

Figure 12 Variation of surface area cleared during diffusion step ($\triangle$A1, ▲), penetration step ($\triangle$A2, □), and rearrangement step ($\triangle$A3, ■) of adsorption of native and succinylated pea legumin at the air–water interface in dependence on the degree of N-modification. (From Ref. 22.)

Characteristic differences were observed for the adsorption isotherms ($\Pi_e = f(\ln c)$) of native and succinylated cruciferin [21] and legumin [22] (Fig. 13). The first part, in which the equilibrium interfacial pressure (Π_e) linearly increased with $\ln c$, has a significantly lower slope, but higher Π_e-values, for both succinylated proteins than for the native one. Since this slope can be related to the surface area covered by one molecule of protein [93,95,96], this parameter can be calculated and serves for the characterization of the surface behavior of the native and succinylated proteins. In both cases, cruciferin and legumin, the exhaustively succinylated form covered a significantly higher interface area [21,22]. Moreover, the succinylated legumin reached the plateau region, which corresponds to the critical micelle concentration of low molecular compounds and may be defined as the critical association concentration of proteins, at lower protein concentration in the bulk phase than the native protein did

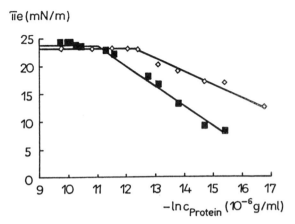

Figure 13 Isotherms of surface adsorption (plot of surface pressure Π_e vs. ln $c_{protein}$) of pea legumin (■) and highly succinylated legumin (◇). (From Ref. 22.)

(Fig. 13). These results indicate an overall higher interfacial activity of the succinylated proteins.

The phenomenon of a higher surface area can be explained by a spread state of the modified protein molecules kept in a rather rigid (less flexible) conformation by the repulsive effects of a great number of succinyl residues in the polypeptide chains. This becomes convincing when the distribution of the potential sites of fixation of succinyl residues (lysine, threonine, serine) over the polypeptide chains is considered. Figure 14 shows the amino acid sequences of the α- and β-chains of legumin. The reactive amino acid residues, bearing succinyl groups when the protein is excessively modified, are uniformly distributed over both polypeptide chains and should exhibit the expected repulsion effect.

Though this interpretation of adsorption kinetics and pressure isotherms is very helpful in deriving structure–functionality relationships, the results are limited to the validity of the theoretical models or approaches. This concerns both the three-step concept of adsorption kinetics and the commonly used simplification of the original Gibbs equation [95], on which the determination of the surface area

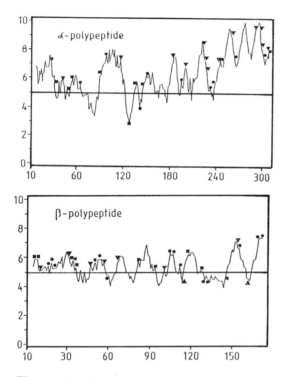

Figure 14 Distribution of the potential sites of fixation of succinyl residues (lysine, ▼; serine, ●; threonine, ■) along the amino acid sequence of the constituent α- and β-polypeptide chains of pea legumin (from Ref. 97) and hydrophobicity profiles determined according to Ref. 98. (From Ref. 22.)

coverage from the isotherms is based. Finding a more sophisticated approach to optimal structure models describing the state of proteins during adsorption and at the adsorbed interface remains a task of high priority in the research of the surface activity of proteins. Chemical modification is a valuable tool in solving this problem.

REFERENCES

1. H. Fraenkel-Conrat, R. S. Bean, and H. Lineweaver, *J. Biol. Chem.* *177*: 385 (1949).

2. P. H. Maurer and H. Lebovitz, *J. Immunol. 76*: 335 (1956).
3. A. S. F. A. Habeeb, H. G. Cassidy, and S. J. Singer, *Biochim. Biophys. Acta 29*: 587 (1958).
4. P. J. G. Butler, J. I. Harris, B. S. Hartley, and R. Leberman, *Biochem. J. 103*: 78 B (1967).
5. P. J. G. Butler, J. I. Harris, B. S. Hartley, and R. Leberman, *Biochem. J. 112*: 679 (1969).
6. H. B. F. Dixon and R. N. Perham, *Biochem. J. 109*: 312 (1968).
7. H. Oppenheimer, K. Bárány, G. Hamor, and J. Fenton, *Arch. Biochem. Biophys. 115*: 233 (1966).
8. H. Oppenheimer, K. Bárány, G. Hamor, and J. Fenton, *Arch. Biochem. Biophys. 120*: 108 (1967).
9. S. K. Gandhi, J. R. Schultz, F. W. Boughey, and R. H. Forsythe, *J. Food Sci. 33*: 163 (1968).
10. H. S. Groninger, *J. Agric. Food Chem. 21*: 978 (1973).
11. L.-F. Chen, T. Richardson, and C. H. Amundson, *J. Milk Food Technol. 38*: 89 (1975).
12. K. L. Franzen and J. E. Kinsella, *J. Agric Food Chem. 24*: 788 (1976).
13. L. R. Beuchat, *J. Agric. Food Chem. 25*: 258 (1977).
14. Y. Sato and R. Nakamura, *Agric. Biol. Chem. 41*: 2163 (1977).
15. H. Aoki, N. Orimo, R. Shimazu, and K. Wakabayashi, *Nippon Shokuhin Kogyo Gakhaishi 25*: 668 (1978).
16. M. Canella, G. Castriotta, and A. Bernardi, *Lebensm.-wiss. u. Technol. 12*: 95 (1979).
17. Y. R. Choi, E. W. Lusas, and K. C. Rhee, *J. Food Sci. 46*: 954 (1981).
18. E. J. Rauschal, K.-J. Linow, W. Pähtz, and K. D. Schwenke, *Nahrung 25*: 241 (1981).
19. K. J. Barber and J. J. Warthesen, *J. Agric. Food Chem. 30*: 930 (1982).
20. P. W. Gossett and R. C. Baker, *J. Food Sci. 48*: 1391, 1395 (1983).
21. J. Gueguen, S. Bollecker, K. D. Schwenke, and B. Raab, *J. Agric. Food Chem. 38*: 61 (1990).
22. M. Subirade, J. Gueguen, and K. D. Schwenke, *J. Colloid Interface Sci. 152*: 442 (1992).
23. G. E. Means and R. E. Feeney, *Chemical Modification of Proteins*, Holden-Day, San Francisco, 1971.
24. A. D. Gounaris and G. E. Perlman, *J. Biol. Chem. 242*: 2739 (1967).
25. T. A. Eisele and C. J. Brekke, *J. Food Sci. 46*: 1095 (1981).

26. P. Plietz and G. Damaschun, *Studia Biophys. 116:* 153 (1986).
27. K. J. Shetty and M. S. N. Rao. *Int. J. Peptide Proteins Res. 11:* 305 (1978).
28. K. D. Schwenke, E. Rauschal, D. Zirwer, and K. J. Linow, *Int. J. Peptide Protein Res. 25:* 347 (1985).
29. K. D. Schwenke, D. Zirwer, K. Gast, E. Görnitz, K.-J. Linow, and J. Gueguen, *Eur. J. Biochem. 194:* 621 (1990).
30. K. D. Schwenke, K.-J. Linow, and D. Zirwer, *Nahrung 30:* 263 (1986).
31. K. D. Schwenke, D. Zirwer, K. Gast, K.-J. Linow, E. Görnitz, J. Gueguen, and M. Subirade, in *Protein Interactions* (H. Visser, ed.), VCH, Weinheim, 1992, pp. 233–252.
32. T. S. Chang and S. F. Sun, *Int. J. Peptide Protein Res. 11:* 65 (1978).
33. J. F. Riordan and B. L. Vallee, *Biochemistry 3:* 1768 (1964).
34. A. Mühlrad, A. Corsi, and A. L. Granata, *Biochim. Biophys. Acta 162:* 435 (1968).
35. M. Siu and U. Thompson, *J. Agric. Food Chem. 30:* 743 (1982).
36. A. F. S. A. Habeeb, *Biochim. Biophys. Acta 121:* 21 (1966).
37. A. F. S. A. Habeeb, *Arch. Biochem. Biophys. 121:* 652 (1967).
38. A. Jonas and G. Weber, *Biochemistry 9:* 4729 (1970).
39. I. M. Klotz and S. Keresztes-Nagy, *Biochemistry 2:* 445 (1963).
40. L. F. Hass, *Biochemistry 3:* 535 (1964).
41. C. L. Polyanowsky, *Biochem. Biophys. Res. Commun. 19:* 364 (1965).
42. R. H. Frist, I. J. Bendet, K. M. Smith, and M. A. Lauffer, *Virology 26:* 558 (1965).
43. W. N. Poillon and A. G. Bearn, *Biochim. Biophys. Acta 127:* 407 (1966).
44. R. Jaenicke and S. Knof, *Eur. J. Biochem. 4:* 157 (1968).
45. R. Jaenicke, D. Schmid, and S. Knof, *Biochemistry 7:* 919, (1968).
46. M. C. Grant-Greene and F. Friedberg, *Int. J. Protein Res. 11:* 235 (1970).
47. A. Spector and E. Katz, *J. Biol. Chem. 240:* 1979 (1965).
48. A. Scanu, W. Reader, and C. Edelstein, *Biochim. Biophys. Acta 160:* 32 (1968).
49. E. A. Meighen and H. K. Schachman, *Biochemistry 9:* 1163 (1970).
50. E. A. Meighen, V. Pigiet, and H. G. Schachman, *Proc. Nat. Acad. Sci.* (Washington) *65:* 234 (1970).
51. S. Keresztes-Nagy, L. Lazer, M. H. Klapper, and I. M. Klotz, *Science 150:* 357 (1965).

52. E. Derbyshire, D. J. Wright, and D. Boulter, *Phytochemistry 15*: 3 (1972).

53. P. Plietz, B. Drescher, and G. Damaschun, *Int. J. Biol. Macromol. 9*: 161 (1987).

54. P. Plietz, G. Damaschun, D. Zirwer, G. Gast, B. Schlesier, and K. D. Schwenke, *Kulturpflanze 32*: S159 (1984).

55. B. Schlesier, *Kulturpflanze 32*: S223 (1984).

56. K. D. Schwenke, R. Mothes, B. Raab, H. Rawel, and J. Gueguen, *Nahrung 37*: 519 (1993).

57. M. A. H. Ismond, E. D. Murray, and S. D. Arntfield, *Int. J. Peptide Protein Res. 26*: 584 (1985).

58. K. D. Schwenke, R. Mothes, D. Zirwer, J. Gueguen, and M. Subirade, in *Food Proteins—Structure and Functionality* (K. D. Schwenke and R. Mothes, eds.), VCH, Weinheim, 1993, pp. 143–153.

59. D. Crenwelge, C. Dill, P. Tybor, and W. Landmann, *J. Food Sci. 39*: 175 (1974).

60. K. Yasumatsu, K. Sawada, S. Moritaka, M. Misaki, J. Toda, T. Wada, and K. Ishii, *Agric. Biol. Chem. 36*: 719 (1972).

61. A. Pearson, M. Spooner, G. Hegarty, and L. Bratzler, *Food Technol. 19*: 1841 (1965).

62. C. Swift and W. Sulzbacher, *Food Technol. 17*: 224 (1963).

63. P. Becher, *Emulsions—Theory and Practice,* Reinhold, New York, 1965, p. 89.

64. C.-Y. Ma, L. M. Poste, and J. Holme, *Can. Inst. Food Sci. Technol. J. 19*: 17 (1986).

65. K. D. Schwenke, E. J. Rauschal, K.-J. Linow, and W. Pähtz, *Nahrung 25*: 201 (1981).

66. C.-Y. Ma, *J. Food Sci. 49*: 1129 (1984).

67. Y. R. Choi, E. W. Lusas, and K. C. Rhee, *J. Amer. Oil Chemists' Soc. 58*: 1044 (1981).

68. E. A. Childs and K. K. Park, *J. Food Sci. 41*: 713 (1976).

69. E. H. Rahma and M. S. N. Rao, *J. Agric. Food Chem. 31*: 352 (1983).

70. M. Kabirullah and R. B. H. Wills, *J. Food Technol. 17*: 235 (1982).

71. K. D. Schwenke and E. J. Rauschal, *Nahrung 27*: 1015 (1983).

72. E. A. Johnson and C. J. Brekke, *J. Food Sci. 48*: 722 (1983).

73. K. Narayana and M. S. N. Rao, *J. Food Sci. 49*: 547 (1984).

74. L. Thompson and Y.-S. Cho, *J. Food Sci. 49*: 1584 (1984).

75. E. Nitecka and K. D. Schwenke, *Nahrung 30*: 969 (1986).

76. A. T. Paulson and M. A. Tung, *J. Food Sci. 53*: 817 (1988).

77. R. Ponnampalam, J. Delisle, Y. Gagné, and J. Amiot, *J. Amer. Oil Chemists' Soc. 67*: 531 (1990).
78. K. L. Franzen and J. E. Kinsella, *J. Agric. Food Chem. 24*: 914 (1976).
79. S. Nakai, *J. Agric. Food Chem. 31*: 672 (1983).
80. S. Nakai, E. Li-Chan, and S. Hayakawa, *Nahrung 30*: 327 (1986).
81. M. L. Crouch, K. M. Tenbarge, A. E. Simon, and R. Ferl, *J. Molec. Appl. Genetics 2*: 273 (1983).
82. M. L. Ericson, J. Rödin, M. Lenman, K. Glimelius, L.-G. Josefsson, and L. Rask, *J. Biol. Chem. 261*: 14576 (1986).
83. E. Nitecka, B. Raab, and K. D. Schwenke, *Nahrung 30*: 975 (1986).
84. K. D. Schwenke, Y. H. Kim, J. Kroll, E. Lange, and G. Mieth, *Nahrung 35*: 293 (1991).
85. J. E. Kinsella, *CRC Crit. Rev. Food Sci. Nutr. 7*: 219 (1976).
86. K. D. Schwenke, E. J. Rauschal, and K. D. Robowsky, *Nahrung 27*: 335 (1983).
87. L. Prahl and K. D. Schwenke, *Nahrung 30*: 311 (1986).
88. L. U. Thompson and E. S. Reyes, *J. Dairy Sci. 63*: 715 (1980).
89. F. MacRitchie and A. E. Alexander, *J. Colloid Sci. 18*: 453 (1963).
90. F. MacRitchie and A. E. Alexander, *J. Colloid Sci. 18*: 458 (1963).
91. D. E. Graham and M. C. Phillips, *J. Colloid Interface Sci. 70*: 403 (1979).
92. A. F. H. Ward and L. Tordai, *Rec. Trav. Chim. Pays-Bas 71*: 572 (1952).
93. E. Tornberg, *J. Colloid Interface Sci. 64*: 391 (1978).
94. E. Tornberg, Y. Granfeld, and C. A. Hakansson, *J. Sci. Food Agric. 33*: 904 (1982).
95. F. MacRitchie, *Adv. Protein Chem. 32*: 283 (1978).
96. D. E. Graham and M. C. Phillips, *J. Colloid Interface Sci. 70*: 415 (1979).
97. G. W. Lycett, R. R. D. Croy, A. H. Shirsat, and D. Boulter, *Nucl. Acids Res. 12*: 4493 (1984).
98. T. R. Hopp and K. R. Wood, *Proc. Natl. Acad. Sci. USA 78*: 3824 (1981).
99. D. W. Sears and S. Beychok, in *Physical Properties and Techniques of Protein Chemistry, Part C* (S. J. Leach, ed.), Academic Press, New York, 1973, pp. 445–593.
100. M. C. Manning and R. W. Woody, *Biochemistry 28*: 8609 (1989).
101. O. B. Ptitsyn, in *Protein Folding* (Th.E. Creighton, ed.), W. H. Freeman, New York, 1992, pp. 243–300.

4
Deamidation and Phosphorylation for Food Protein Modification

Frederick F. Shih

Southern Regional Research Center, USDA, New Orleans, Louisiana

I. INTRODUCTION

Food proteins, especially those of plant origin, often require modification to achieve desirable functional properties for use as food ingredients. For instance, soy protein has limited water solubility at acid pH, which restricts its use in acidic foods such as coffee whitener and acidic beverages. Improved solubility at acid pH for commercial soy protein isolate can generally be achieved by hydrolysis. However, the hydrolysis has to be carefully controlled, because excessive peptide bond hydrolysis may release bitter peptides, resulting in undesirable off-flavors. Scientists are constantly looking for better and safer methods to improve the functional properties of protein to meet the needs of the food industry.

Functional properties of food protein are sensitive to changes in the size of the protein molecule, structural conformation, and the level and distribution of ionic charges. They can be modified, both enzymatically and nonenzymatically, by reactions such as protein hydrolysis, denaturation, ionization, and cross-linking. Enzymatic modifications are considered safer for food uses and therefore more desirable. However, many nonenzymatic methods have been proven safe and, because of their simplicity and great effectiveness, they are often the method of choice. Specifically, food proteins have been modified for improved functional properties by succinylation, acetylation, glycosylation, phosphorylation, and specific amide bond hydrolysis (deamidation). General reviews on the subject up to 1994 are available in the literature [1–5].

In recent years, deamidation and phosphorylation for food protein modification have received considerable attention among food research laboratories. These two methods are popular for good reasons. Firstly, both deamidation–amidation and phosphorylation–dephosphorylation occur naturally in living cells. The deamidated or phosphorylated products, prepared in the laboratory, normally maintain their natural forms and original nutritional values. Secondly, both reactions increase the negative charge density of the protein without significantly decreasing the molecular weight. The modified protein is more hydrophilic and soluble in water because of the increased ionic charge. Other functional properties, such as emulsifying activity, are generally improved because they are strongly correlated to solubility, especially when macromolecular characteristics of the protein remain intact.

In this chapter, a review of protein deamidation and phosphorylation will be given, with emphasis on recent developments relating these reactions to use in foods. The literature on both enzymatic and nonenzymatic modifications will be included. While traditional and popular methods are summarized and remain the focus of this review, new and innovative approaches will be introduced and given special attention.

II. PROTEIN DEAMIDATION

Protein deamidation converts amide groups in glutamine and asparagine residues to acid groups with the concomitant release of ammonia. In nature, protein deamidation plays an important role in biological regulations, probably as a biological clock in determining the function and catabolism of proteins [6] or as a signal for the degradation of enzymes and hormones [7]. In food protein modification, food scientists are just beginning to recognize its potential as a viable alternative to protein hydrolysis. The following review will not cover the biological aspects of protein deamidation. The survey will be restricted to the methodologies and applications of deamidation as available in the literature and useful to the food industry.

A. Nonenzymatic Deamidation

Proteins are most conveniently deamidated by acid or base hydrolysis. The reaction mechanism has been extensively investigated. Figure 1 summarizes the pathways of nonenzymatic deamidation re-

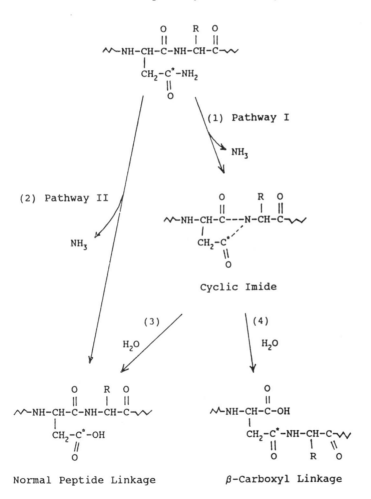

Figure 1 Mechanism of nonenzymatic deamidation.

ported in the literature. The conversion products vary depending on factors including pH, temperature, buffer composition, and sequence and size of amino acids in the deamidated substrate [8–10].

Particularly, pH appears to exert major control over the nonenzymatic deamidation of asparaginyl residues in polypeptides. At alkaline and neutral pH's, deamidation of asparagine (Asn) residues was reported to involve the formation of an intramolecular cyclic imide intermediate (Fig. 1, Pathway I) [7,11,12]. The intermediate could be hydrolyzed to a normal α-linked aspartate (Asp) residue and a β-linked Asp residue (iso-Asp), with the latter normally predominating [13–15]. Under acidic conditions, the reaction was reported to proceed via direct hydrolysis of Asn residues to Asp residues without the formation of a cyclic imide intermediate (Fig. 1, Pathway II) [8]. This straightforward deamidation under acidic conditions is significant because chemical deamidation of food proteins, normally conducted in mildly acidic solutions, is expected to produce the normal α-linked carboxylic residues instead of the nutritionally undesirable β-linked isomers. The ease of deamidation depends both on the amino acid sequence flanking the carboxamide group [12,16] and on the tertiary structure of the protein [17]. While both asparagine and glutamine can undergo deamidation, asparagine has been shown to deamidate at much higher rates [13,18].

1. Mild Acid Catalyzed Deamidation

As mentioned earlier, deamidation by excessive protein hydrolysis is undesirable for food use because it could result in reduced protein functionality and the release of bitter-tasting peptides [19]. Alkali-catalyzed deamidation is also undesirable because, in addition to indiscriminate hydrolysis, alkali treatment results in products such as lysinoalanine that have been implicated in kidney damage in rats [20,21].

Therefore most nonenzymatic deamidations for food protein modification are conducted under mild acid and relatively low temperature conditions. For instance, according to Wu and coworkers [22], the best conditions to produce soluble gluten was to deamidate a 5% gluten solution at 121°C either in 0.02 N HCl for 30 min or in 0.05 N HCl for 15 min. However, optimum conditions should be determined for each individual case. Ideally, significant deamidation is achieved to ensure maximum improvements in functionality for the protein, while relatively low and insignificant peptide bond hydrolysis is maintained to avoid destroying the macromolecular structure of the protein.

For plant proteins, such as soy protein and wheat gluten, which are rich in amide amino acids, even low-level 2–5% deamidation could result in significant improvement of their functional properties. The mild acid deamidation of soy protein or wheat gluten and its effects on changes in physicochemical properties have been investigated extensively [23–26]. In a typical example, Matsudomi et al. [23] treated soy protein with 0.05 N HC1 at 95°C for 90 min, resulting in up to 30% deamidation and about 10% peptide bond hydrolysis (Fig. 2). Solubility increased markedly for the deamidated products in the isolelectric region. Similar increase in surface hydrophobicity was also observed (Fig. 3) [25], indicating that the hydrophobic regions were exposed as a result of deamidation. The authors verified this molecular unfolding by showing a decrease in the

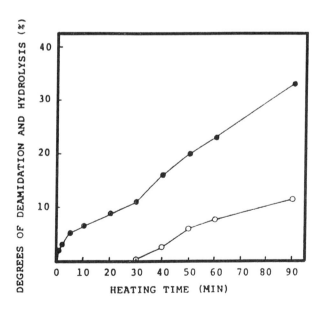

Figure 2   Degree of deamidation (●) and peptide bond hydrolysis (○) of soy protein during mild acid treatment. (From Ref. 23. Reprinted by permission.)

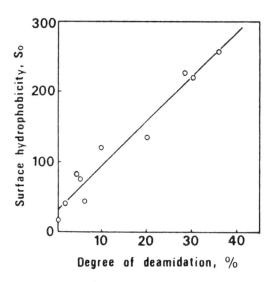

Degree of deamidation, %

Figure 3 Surface hydrophobicity as a function of gluten deamidation. (From Ref. 25. Reprinted by permission.)

helical content of the protein upon deamidation (Fig. 4). Invariably, emulsifying activity, emulsion stability, and other emulsifying properties increased with increased surface hydrophobicity.

Numerous applications for mild acid deamidated proteins have been reported in the literature. MacDonald and Pence [27] investigated the effects of deamidation on foaming characteristics of gliadin and found that the performance of gliadin deamidated in mild acid (0.07 N HCl) compared favorably to egg white as a foaming agent in a variety of food applications. Finley [28] treated gluten with 0.5 N HCl at 95°C for up to 30 min and found that the deamidated gluten contained 13.9% nitrogen, of which 22.4% was amide nitrogen. Gluten thus modified, with a relatively low level (5–20%) of deamidation, showed excellent dispersibility and could be utilized as a fortifier for fruit juices.

2. Anion-Catalyzed Deamidation

Various methods have been developed to enhance deamidation selectively with minimum peptide bond hydrolysis. It has been known

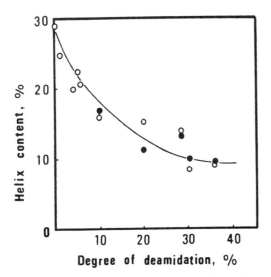

Figure 4 Helix content as a function of gluten deamidation. (From Ref. 25. Reprinted by permission.)

that, in the presence of anions, particularly those with long carbon chains, such as alkylsulfates and arylsulfonates, the hydrolysis of polypeptide and protein is generally accelerated [29,30]. Later reports [31–34] demonstrated that, under relatively mild acid and temperature conditions, the anion-catalyzed enhancement was more extensive for deamidation than for peptide bond hydrolysis. Soy proteins were successfully deamidated by 0.1–0.2 N HCl at 75–85°C in the presence of small amounts of dodecylsulfate or arylsulfonates in the form of cation exchange resins [31]. The modified proteins, with up to 40% deamidation and only 1–4% peptide bond hydrolysis, showed substantially improved solubility, water binding capacity, foam expansion, and emulsion activity over the untreated control. The process using cation exchange resins is particularly cost-effective and amenable to industrial scale-up [34].

3. Extrusion-Enhanced Deamidation

Another novel approach to achieve effective protein deamidation is by extrusion. Extrusion technology provides, in addition to the compounding effect, a high-temperature, short-time (HTST) environment in the extruder. While prolonged heating normally results in indiscriminate protein hydrolysis, HTST should, in theory, favor the hydrolysis of the weaker amide bond, or deamidation, over the peptide bond hydrolysis. Indeed, extrusion has been found to be extremely effective in enhancing protein deamidation [35]. When gluten was extruded at temperatures from 130 to 170°C, deamidation increased markedly with increased temperature. The authors theorized that the rate increase at elevated temperature became more pronounced with extrusion than without because the extrusion caused unfolding of the protein structure, which led to easy penetration for water molecules and ready accessibility of the reaction sites. For the same reason, when feed moisture was increased from 20 to 30% at sufficiently high temperatures, the deamidation rate also increased.

4. Maillard-Related Deamidation

Protein deamidation has been utilized in the development of specific functionality for a variety of protein products. Of particular interest is the implication that deamidation plays a role in the generation of color, aroma, and flavor in heated foods [36]. Both ammonia and amide amino acids in heated model systems were reported to correlate with the formation of the flavor compound pyrazines, which are normally produced in the Maillard reaction [37,38]. Ammonia released during deamidation appeared to be a key element in the flavor-generating mechanism [39,40]. When gluten at different levels of deamidation was heated with glucose in complex model systems, the generation of Maillard compounds decreased with increased deamidation, demonstrating again the positive effect of ammonia on the Maillard reaction. The Maillard compounds identified included various furans, which are known to provide pleasant flavor and aroma to heated foods [36]. This research contributes to the understanding of the complex Maillard system and may eventually lead to the development of flavor in foods by protein deamidation under controlled Maillard reaction conditions.

B. Enzymatic Deamidation

Protein deamidation using enzymes is generally more desirable than using chemicals because, in addition to being safer, enzymes are substrate specific, that is, no peptide bond hydrolysis occurs. Enzyme reactions are also catalyzed under mild temperature (30–40°C) and neutral pH conditions to keep undesirable products normally associated with harsh reaction conditions to a minimum. However, only a few deamidating enzymes are available, and methods for enzymatic deamidation are yet to be developed for practical large-scale applications.

1. Deamidation by Peptidoglutaminases

Peptidoglutaminases (PGase) are the only well-characterized L-glutamine amido hydrolase (EC 3.5.1.2) capable of specifically hydrolyzing glutamine to glutamate. Prepared from the soil bacterium Bacillus circulans, the deamidating enzymes consisted of PGase I and PGase II, which catalyzed the deamidation of C-terminal glutamine in small peptides and glutamine within the peptide chain respectively [41]. According to Hamada et al. [42], a preparation of PGases readily deamidated glutamine residues in soy protein hydrolysate, but its activity toward intact soy protein was small. The use of proteolysis or heat treatment prior to PGase treatment was found to be effective in enhancing the deamidation [43].

2. Deamidation by Deamidases

Shutov and Vaintraub [44] reported intensive deamidation activities of protein extracts from germinating seeds. Protein deamidases, prepared and purified from germinating wheat, kidney bean, and squash, were found to be specific to glutamine deamidation, not affected by reducing reagents, and partially inhibited by calcium ions [45]. Relative activity of these enzymes toward a variety of protein substrates were reported (Table 1) [46].

3. Deamidation by Transglutaminase

Transglutaminase (TGase, EC 2.3.2.13) catalyzes the incorporation of primary amines into proteins and polypeptides through an acyl transfer reaction between the γ-carbonyl group in glutamine and a receptor amino group as [47]

$$[PROTEIN]\text{-}CONH_2 + R\text{-}NH_2 \rightarrow [PROTEIN]\text{-}CONHR + NH_3$$

Normally, the amino group is part of the same or another protein, and the TGase catalysis results in intra- or intermolecular cross-linking.

Table 1 Relative Activity of Deamidases from Various Seed Sources to Proteins, Peptides, and Amino Acids

	Relative activity, %[a]		
Substrate	Wheat	Kidney bean	Squash
Soybean globulin	100	100	100
Soybean 11S	47	54	58
Squash 11S	67	77	76
Soybean 7S	90	48	49
Kidney bean 7S	90	56	51
Kidney bean hemagglutinin	35	34	37
Wheat gluten	48	43	49
Horse hemoglobin	9	18	11
Lysozyme	13	19	32
Reduced RNAase	55	24	28
Glutamine	25	25	14
Gly-Gln-Tyr	75	—	—

[a]Normalized relative to the activity to soybean globulin.
Source: Vaintraub et al., Ref. 46. Reprinted by permission.

However, if no free amino groups are present, or if the amino groups are blocked, water can act as an acyl acceptor, and deamidation occurs with the release of ammonia to give peptide-bound glutamic acid.

$$[\text{PROTEIN}]\text{-CONH}_2 + \text{H}_2\text{O} \rightarrow [\text{PROTEIN}]\text{-COOH} + \text{NH}_3$$

Therefore TGase is basically a cross-linking enzyme and often used as such in food protein modification [48,49]. However, under controlled conditions, TGase can be used to catalyze deamidation predominantly. Prior to TGase catalysis, proteins have been acylated, deaminated, or guanidinated to avoid cross-linking [50,51]. If proteins are acylated with citraconic anhydride, the blockage is reversible and the citroconylated residue can be regenerated under acidic conditions [52]. As an example, when Motoki et al. used TGase to catalyze the deamidation of α_{s1}-casein, citraconylation was performed prior to deamidation [53]. Almost complete citraconylation

Native α_{s1}-casein (pH$_{iso}$ 4.9)

↓ Citraconic anhydride (pH 8.0, 6M urea)

Citraconylated α_{s1}-casein (pH$_{iso}$ 3.9, 96%)

↓ Transglutaminase (0.02 U/mg, pH 7.4)

Deamidated and citraconylated α_{s1}-casein (79%)

↓ HCl (pH 3.3, 4 h, 37°C)

Deamidated α_{s1}-casein (pH$_{iso}$ 4.4, 103%)

Figure 5 A flow diagram showing the deamidation of α_{1s}-casein by transglutaminase. pH$_{iso}$ refers to the protein isoelectric pH, and the percentages indicate the degree of modification. (From Ref. 53. Reprinted by permission.)

was observed and about 80% of the glutamine residues were deamidated (Fig. 5). After removal of the blocking groups, the deamidated protein was found to have improved solubility, particularly in the pH range between 3 and 5.

4. Deamidation by Proteases

Kato et al. [54,55] reported using the proteases papain, pronase E, and chymotrypsin for the deamidation of food proteins, and Kato et al. [56] later reported deamidation of selected plant and animal proteins with chymotrypsin immobilized on controlled pore glass. The reactions were carried out at alkaline pH (pH 10) and 20°C and resulted in 5 to 20% deamidation with up to 8% peptide bond hydrolysis. Functional properties of proteins thus deamidated showed increased solubility and emulsifying and foaming properties. Enzymatic deamidation using commercial food-grade proteases appears to be a straightforward and cost-effective process for food protein modification. However, the deamidating activity of the enzymes was nonspecific, because both amide and peptide bond hydrolysis occurred. Also, using the method of Kato et al. [54], both Bollecker et al. [57] and Shih [58] could not confirm the deamidating activity of the protease. According to Shih [32,58], the deamidation may or may not be enzymatic, because the ammonia generated could be from anion-catalyzed deamidation of free amino acids in the hydrolysate. Further research is needed to explore the exact role of proteases in protein deamidation.

III. PROTEIN PHOSPHORYLATION

Inorganic phosphate (Pi) can be transferred to proteins by either O- or N-esterification reactions. In O-esterification, Pi reacts with the primary or secondary hydroxyl on serine or threonine respectively, or with the weakly acidic hydroxyl on tyrosine, forming a -C-O-Pi bond. In N-esterification, Pi combines with the γ-amino group of lysine, the imidazole group of histidine, or the guanidino group of arginine, forming a -C-N-Pi bond. The N-bound phsphates are acid labile and are readily hydrolyzed below pH 7 (59). Proteins containing O-bound phosphates are acid stable and are the modification of choice for food proteins [60], since the pH of most foods is 3–7.

A. Nonenzymatic Phosphorylation

Reviews on methods for nonenzymatic phosphorylation of food proteins are available, covering the literature up to 1991 [60–62]. Of the chemical reagents surveyed, phosphorus oxychloride ($POCl_3$) and sodium trimetaphosphate (STMP) have been investigated extensively. However, only $POCl_3$ appears to be suitable and ready for large-scale application. Food proteins frequently phosphorylated include soy protein, wheat gluten, and yeast protein.

1. Phosphorylation by Phosphorus Oxychloride

Generally, $POCl_3$ reacts with water generating active intermediates that phosphorylate proteins to yield either N-bound or O-bound phos-

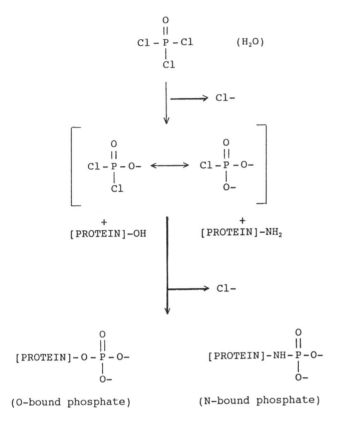

Figure 6 Reaction diagram of protein phosphorylation using $POCl_3$.

phates (Fig. 6). To minimize the excessive heat produced from the reaction with water, $POCl_3$ is usually dissolved in an organic solvent such as carbon tetrachloride and added in small portions to aqueous protein solutions. The pH is kept between 5 and 9 with sodium hydroxide, and the temperature is controlled in an ice bath.

Hirotsuka et al. [63] phosphorylated lysine and histidine residues on soy protein isolate with $POCl_3$ and found the phosphorylated protein to be stable above pH 3.5 and have increased solubility and emulsifying activity. Similarly using $POCl_3$, Woo and Richardson [64] observed increased emulsifying activity and gel forming properties in the presence of calcium ions for the phosphorylated β-lacto-globulin. In contrast, both $POCl_3$-phosphorylated casein [59] and β-lactoglobulin [65] were found to have decreased water solubility and emulsifying activity due to protein cross-linking. Cross-linking could also be the cause for mixed results of increased solubility but decreased emulsifying activity for the $POCl_3$-phosphorylated zein hydrolysate [66].

The high content of nucleic acid in yeast is a potential problem associated with the consumption of large amounts of yeast nucleoprotein in foods. Phosphorylation is one of the methods used for reducing nucleic acid of yeast proteins (Table 2) [67]. Huang and Kinsella [68] reported the effective removal of nucleic acid from yeast proteins by $POCl_3$-phosphorylation. The phosphorylated protein showed improvement in emulsifying activity and emulsion stability, and produced stable but weak foams at neutral pH. The authors also noted that the in vitro digestibility of yeast protein was not affected by phosphorylation as reported for casein [60] and soy protein [65].

In $POCl_3$-phosphorylated food proteins, Hirotsuka et al. [63] reported up to 140 moles of Pi incorporated per mole of soy isolate, while Woo et al. [65] reported 14 moles of Pi per mole of β-lacto-globulin, and Matheis et al. [59] found 7.4 and 6.2 moles of Pi per mole of casein and lysozyme respectively. $POCl_3$ is an effective phosphorylating agent; however, it reacts rapidly with water to produce Pi and the noxious gas HCl. It also promotes protein cross-

Table 2 Proximate Composition (%) of Yeast, Yeast Nucleoprotein, and Yeast Proteins Prepared by Different Methods

Sample	Protein	Nucleic acid	Carbohydrate	Lipid
Whole yeast cell	42.4	11.4	28.8	4.0
Nucleoprotein	64.0	20.0	6.0	5.1
Succinylated protein	90.0	1.8	3.0	3.5
Citraconylated protein	84.0	2.2	3.7	5.1
Phosphorylated protein	82.0	2.7	5.1	5.7

Source: Kinsella, Ref. 67. Reprinted by permission.

linking, and its site of phosphorylation is nonspecific, as both O- and N-esterification occur.

2. Phosphorylation by Trimetaphosphate

Sung et al. [69] used sodium trimetaphosphate (STMP) to modify serine and lysine in soy isolate under alkaline conditions. About 40% of the total serine residues were phosphorylated with no protein cross-linking, and the isolate displayed increased solubility and emulsifying properties, particularly under acidic conditions. Giec et al. [70] also reported the phosphorylation of a yeast homogenate using STMP to remove substantially the nucleic acid and prepare a protein isolate with markedly improved functional properties including solubility and emulsifying properties. However, Matheis et al. [59] were unable to detect any covalently bound Pi in soy protein and lysozyme using the methods of Sung et al. [69].

Nonenzymatic phosphorylation with STMP appears to be able to solve many of the problems associated with $POCl_3$. For example, STMP is an FDA-approved food additive [71], does not cause protein cross-linking, and its hydrolysis in water produces only harmless Pi. However, phosphorylation with STMP occurs at alkaline pH, which could lead to undesirable reactions and products as described earlier. STMP shows potential for food protein modification, but further research is needed to determine the exact incorporation of covalently bound Pi, to avoid alkaline pH conditions, and to move its specificity toward O-esterification rather than a mixture of O- and N-esterifications.

B. Enzymatic Phosphorylation

Enzymatic phosphorylation has been extensively investigated because protein phosphorylation–dephosphorylation is an important mechanism in the regulation of a variety of enzymes and proteins in mammalian cells [72–75]. However, very few reports are available that describe enzymatic phosphorylation for food application. Enzymes that phosphorylate proteins are protein kinases (EC 2.7.1.37), which transfer the γ-phosphate of ATP to the hydroxyl groups of serine, threonine, or tyrosine residues in proteins. Numerous protein kinases have been identified and characterized, but none are available in quantities large enough for practical food use.

1. Protein Kinase from Bovine Cardiac Muscle

The protein kinase with the most potential for food use is probably the adenosine cyclic $3',5'$-monophosphate–dependent protein kinase (cAMPdPK) from bovine cardiac muscle [76,77]. The amino acid sequence of the enzyme has been determined [78,79]. The DNA segment that encodes this kinase has been isolated, cloned, and expressed [80–82]. Thus large quantities of active enzyme could be made economically available if it proves useful in food protein modification.

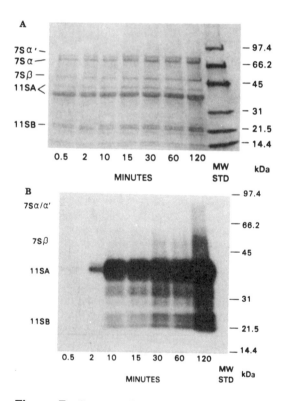

Figure 7 Soy protein phosphorylation using cAMPdPK as a function of time: (A) electrophoretic profile of the protein subunits; (B) autoradiogram of the electrophoretic profile. (From Ref. 85. Reprinted by permission.)

The phosphorylating sites of cAMPdPK and the cAMPdPK phosphorylation of soy proteins have been intensively investigated. The enzyme is most active toward serines that are located one or two amino acid residues away from basic amino acids such as arginine and lysine, and the highest specificity is toward the sequences -Arg-Arg-X-Ser- and -Lys-Arg-X-Ser-, where X represents any amino acid [73]. There are at least 18 phosphorylation sites on the acidic polypeptides of soy glycinin (11S subunits), one on β-conglycinin (7S subunits), but none on the basic polypeptides of glycinin [83,84]. The ease of phosphorylation using cAMPdPK for soy protein subunits was studied in a timed sequence by scanning the electrophoretic profile of the products in an autoradiogram as shown in Fig. 7 [85]. As expected, Pi was preferentially incorporated into the acidic over the basic glycinin subunits, and glycinin was invariably phosphorylated to a greater extent than β-conglycinin. However, under optimum conditions, no more than 25% of the known phosphorylation sites in glycinin were phosphorylated [84]. As shown in Table 3, the effectiveness of soy protein phosphorylation ranged from 0.2 to 6

Table 3 Moles of Phosphorus Incorporated per Mole of Soy Protein Using Different cAMPdPK Preparations

	β-Conglycinin		Glycinin	
Ultrafiltrate isolate				
native	0.16[a]	(1.15)[b]	0.70[a]	(1.27)[b]
denatured	0.83	(0.66)	1.01	(1.38)
Purina Protein 620				
native	0.45	(1.16)	1.14	(1.36)
denatured	0.88	(0.83)	0.53	(2.09)
β-conglycinin				
native	0.28	(0.04)		
denatured	1.20	(0.21)		
Glycinin				
native			0.63	(2.08)
denatured			3.96	(5.57)

[a] Pi incorporation using laboratory-prepared cAMPdPK.
[b] Pi incorporation using commercial cAMPdPK.
Source: Ross and Bhatnagar, Ref. 86; Ross, Ref. 87. Reprinted by permission.

moles Pi/mole protein, depending on the purity of the enzyme and the protein substrate [86,87]. Even with this low level of phosphorylation, the modified soy proteins often showed significant improvement in solubility and emulsifying properties [84,85]. Further research will enhance the efficiency of the cAMPdPK-catalyzed phosphorylation, and greater impact is expected for the modification on the improvement of food protein qualities.

2. Casein Kinase from Mammary Gland

Casein kinases exist as two distinct sets of enzymes [88]. The casein kinase in mammary tissue that normally provides in vivo phosphorylation of nascent proteins is distinct from the multisubstrate and ubiquitous casein kinase (CK-2) that is responsible for in vitro casein phosphorylation. The substrate specificity of the two enzymes is also different. For the casein kinase from mammary gland, the recognition sequence corresponds to the tripeptide Ser/Thr-X-Glu/Ser-P or Ser/Thr-X-Glu/Ser in nonphosphorylated proteins, where X is any amino acid [89]. For CK-2, the recognition sites have been identified as Ser-Glu-Ala-Glu-Glu-Glu and Ser-Ala-Ala-Glu-Glu-Glu [88].

Table 4 Phosphorylation of Proteins by a Casein Kinase from Rat Mammary Gland

Protein substrate	Rate of phosphate addition (nmol/mg/20 min)
α_{s1}-Casein	8.6
Dephosphorylated α_{s1}-casein	43.0
β-Casein	2.5
Dephosphorylated β-casein	32.4
κ-Casein	1.3
Dephosphorylated κ-casein	8.5
β-lactoglobulin	2.1
α-lactalbumin	1.1
Fat globule membrane proteins	6.1
Histone-arginine rich	0.1
Histone-lysine rich	0.1
Phosvitin	0.3
Lysozyme	0.0

Source: Bingham and Farrell, Ref. 90. Reprinted by permission.

Besides its in vivo activity, mammary gland casein kinase can phosphorylate proteins other than casein, as demonstrated by the in vitro phosphorylation of several food proteins as shown in Table 4 [90]. As expected, there is a marked preference of the enzyme for dephosphorylated caseins over native casein, and for native caseins over other food proteins. So far, no protein kinases have been evaluated for the phosphorylation of plant proteins. An examination of known primary structural sequences of soy glycinin and β-conglycinin revealed several potential phosphorylation sites for the mammary gland kinase but not for the CK-2 (unpublished observations). Based on this information, mammary gland kinase appears most likely to be active toward soy proteins.

IV. CONCLUSION

This review has surveyed the information available in the literature for the deamidation or phosphorylation of food proteins. The focus is on methodology and application that may be useful for the food industry. Overall, the practical utilization of protein deamidation and phosphorylation remains in the early stages of exploration. Most examples given, other than maybe mild acid deamidation, are still waiting to join the main stream of food protein modification. Some of them, as in the case of $POCl_3$ or cAMPdPK phosphorylation, are well characterized in the laboratory but yet to be evaluated for practical scale-up. A few others have long been in existence but are only recently being recognized as viable modifications, such as the Maillard-related deamidation and extrusion deamidation. Given the ingenuity of scientists working to develop these modifications and what they have accomplished so far, deamidation and phosphorylation are expected to have a significant impact for food protein modification in the future.

REFERENCES

1. R. E. Feeney and J. R. Whitaker, in *New Protein Foods*, Vol. 5 (A. M. Altschul and H. L. Wilcke, eds.), Academic Press, New York, 1985, pp. 181–219.
2. F. F. Shih, in *Biochemistry of Food Proteins* (B. J. F. Hudson, ed.), Elsevier Applied Science, London, 1992, pp. 235–248.

3. J. S. Hamada, in *Biochemistry of Food Proteins* (B. J. F. Hudson, ed.), Elsevier Applied Science, London, 1992, pp. 249–270.
4. J. Leman, *Przemysl-Spozywczy 47*: 212 (1993).
5. J. S. Hamada, *CRC Crit. Rev. Food Sci. & Nutr. 34*: 383 (1994).
6. A. B. Robinson and C. J. Rudd, in *Current Topics in Cellular Regulations*, Vol. 8 (B. L. Horecker and E. R. Stadtman, eds.), Academic Press, New York, 1974, pp. 274–295.
7. D. W. Aswad, *J. Biol. Chem. 259*: 714 (1984).
8. N. P. Bhatt, K. Patel, and R. T. Borchardt, *Pharm. Res. 7*: 593 (1990).
9. K. Patel and R. T. Borchardt, *Pharm. Res. 7*: 703 (1990).
10. K. Patel and R. T. Borchardt, *Pharm. Res. 7*: 787 (1990).
11. Y. C. Meinwald, E. R. Stinson, and A. Scheraga, *Agric. Biol. Chem. 49*: 1251 (1985).
12. T. Geiger and S. Clarke, *J. Biol. Chem. 262*: 785 (1987).
13. E. Sondheimer and R. W. Holley, *J. Am. Chem. Soc. 76*: 2467 (1954).
14. R. Battersby and J. C. Robinson, *J. Chem. Soc. 1955*: 259 (1955).
15. R. Lura and V. Schirch, *Biochemistry 27*: 7671 (1988).
16. J. J. Harding, *Adv. Protein Chem. 37*: 247 (1985).
17. A. A. Kossiakoff, *Science 240*: 191 (1988).
18. A. B. Robinson, J. W. Scotchler, and J. H. McKerrow, *J. Am. Chem. Soc. 95*: 8156 (1973).
19. J. Adler-Nissen, in *Enzymic Hydrolysis of Food Proteins*, Elsevier Applied Science, New York, 1986, pp. 324–329.
20. J. R. Whitaker and R. E. Feeney, *CRC Crit. Rev. Food Sci. Nutr. 19*: 3 (1983).
21. J. C. Woodward and C. D. Short, *J. Nutr. 103*: 569 (1973).
22. C. H. Wu, N. Shuryo, and W. D. Powrie, *J. Agric. Food Chem. 24*: 504 (1976).
23. N. Matsudomi, T. Sasaki, A. Kato, and K. Kobayashi, *Agric. Biol. Chem. 49*: 1251 (1985).
24. A. Kato, N. Tsutsui, K. Matsudomi, K. Kobayashi, and S. Nakai, *Agric. Biol. Chem. 45*: 2755 (1981).
25. N. Matsudomi, A. Kato, and K. Kobayashi, *Agric. Biol. Chem. 46*: 1583 (1982).
26. A. Kato and S. Nakai, *Biochim. Biophys. Acta 642*: 13 (1980).
27. C. E. MacDonald and J. W. Pence, *Food Technol. 15*: 141 (1961).
28. J. W. Finley, *J. Food Sci. 40*: 1283 (1975).

29. J. C. Paulson, F. E. Deatherage, and E. F. Almy, *J. Am. Chem. Soc.* 75: 2039 (1953).
30. J. R. Whitaker and F. E. Deatherage, *J. Am. Chem. Soc.* 77: 3360 (1955).
31. F. F. Shih and A. D. Kalmar, *J. Agric. Food Chem.* 35: 672 (1987).
32. F. F. Shih, *J. Food Sci.* 56: 452 (1991).
33. F. F. Shih, *J. Food Sci.* 52: 1529 (1987).
34. F. F. Shih, U.S. patent 4,824,940 (1989).
35. H. V. Izzo, M. D. Lincoln, and C.-Y. Ho, *J. Agric. Food Chem.* 41: 199 (1993).
36. H. V. Izzo, Ph.D. dissertation, Rutgers University, 1993.
37. H. V. Isso and C.-T. Ho, *J. Agric. Food Chem.* 41: 2364 (1993).
38. H. I. Hwang, T. G. Hartman, and R.-T. Ho, *J. Agric. Food Chem.* 41: 2112 (1993).
39. P. S. Wang, H. Kato, and M. Fugimaki, *Agric. Biol. Chem.* 33: 1775 (1969).
40. P. E. Koehler, M. E. Mason, and J. A. Newell, *J. Agric. Food Chem.* 17: 393 (1969).
41. M. Kikuchi, H. Hayashida, E. Nakano, and K. Sakahuchi, *Biochemistry 10*: 1222 (1971).
42. J. S. Hamada, F. F. Shih, A. W. Frank, and W. E. Marshall, *J. Food Sci.* 53: 671 (1988).
43. J. S. Hamada. *J. Am. Oil Chem. Soc.* 68: 459 (1991).
44. A. D. Shutov and I. A. Vaintraub, *Phytochem.* 26: 1557 (1987).
45. I. A. Vaintraub, L. V. Kotova, and R. Shaha, *FEBS Lett.* 302: 169 (1992).
46. I. A. Vaintraub, L. V. Kotova, and R. Shaha, in *Food Proteins* (K. D. Schwenke and R. Mothes, eds.), VCH, Weinheim, Germany, 1993, pp. 187–189.
47. L. Lorand and S. Conrad, *Mol. Cell. Biochem.* 58: 9 (1984).
48. R. Mahmond and P. A. Savello, *J. Dairy Sci.* 76: 29 (1993).
49. C. Larre, M. Chiarello, Y. Blanloeil, M. Chenu, and J. Gueguen, *J. Food Biochem.* 17: 267 (1994).
50. M. J. Mycek and H. Waelsch, *J. Biol. Chem.* 235: 1357 (1960).
51. D. Bercovici, H. F. Gaertner, and A. J. Puigserver, *J. Agric. Food Chem.* 35: 301 (1987).
52. A. C. Brinegar and J. E. Kinsella, *J. Agric. Food Chem.* 28: 818 (1980).
53. M. Motoki, K. Seguro, N. Nio, and K. Takinami, *Agric. Biol. Chem.* 50: 3025 (1986).

54. A. Kato, A. Tanaka, Y. Lee, N. Matsudomi, and K. Kobayashi, *J. Agric. Food Chem. 35*: 285 (1987).

55. A. Kato, A. Tanaka, N. Matsudomi, and K. Kobayashi, *J. Agric. Food Chem. 35*: 224 (1987).

56. A. Kato, Y. Lee, and K. Kobayashi, *J. Food Sci. 54*: 1345 (1989).

57. S. Bollecker, G. Viroben, Y. Popineau, and J. Gueguen, *Sci. Aliments 10*: 343 (1990).

58. F. F. Shih, *J. Food Sci. 55*: 127 (1990).

59. G. Matheis, M. H. Penner, R. E. Feeney, and J. R. Whitaker, *J. Agric. Food Chem. 31*: 379 (1983).

60. G. Matheis and J. R. Whitaker, *J. Agric. Food Chem. 32*: 699 (1984).

61. A. W. Frank, *Phosphorus and Sulfur 29*: 297 (1987).

62. G. Matheis, *Food Chem. 39*: 13 (1991).

63. M. Hirotsuka, H. Taniguchi, H. Narita, and M. Kito, *Agric. Biol. Chem. 48*: 93 (1984).

64. S. L. Woo and T. Richardson, *J. Dairy Sci. 66*: 984 (1983).

65. S. L. Woo, L. K. Creamer, and T. Richardson, *J. Agric. Food Chem. 30*: 65 (1982).

66. M. L. A. Casella and J. R. Whitaker, *J. Food Biochem. 14*: 453 (1990).

67. J. E. Kinsella, in *Food Biochemistry* (D. Knorr, ed.), Marcel Dekker, New York, 1987, pp. 363–391.

68. Y. T. Huang and J. E. Kinsella, *J. Food Sci. 52*: 1684 (1987).

69. H. Sung, H. Chen, T. Liu, and J. Su, *J. Food Sci. 48*: 716 (1983).

70. A. Giec, B. Stasinska, and J. Skupin, *Food Chemistry 31*: 279 (1989).

71. R. H. Ellinger, in *CRC Handbook of Food Additives*, 2d ed., Vol. 1 (T. E. Furia, ed.), CRC Press, New York, 1972, pp. 640.

72. E. G. Krebs, in *The Enzymes*, Vol. 17 (P. D. Boyer and E. G. Krebs, eds.), Academic Press, New York, 1986, pp. 3–20.

73. E. G. Krebs and J. A. Beavo, *Ann. Rev. Biochem. 48*: 923 (1979).

74. P. Cohen, *Nature 296*: 613 (1982).

75. E. J. Nestler, S. I. Walaas, and P. Greengard, *Science 225*: 1357 (1984).

76. S. B. Smith, J. B. White, J. B. Siegel, and E. G. Krebs, in *Protein Phosphorylation* (O. R. Rosen and E. G. Krebs, eds.), Cold Spring Harbor Laboratory, Cold Spring Harbor, ME, 1981, pp. 55–56.

77. S. Okuno and H. Fujisawa, *Biochim. Biophys. Acta 1038*: 204 (1990).

78. S. Shoji, D. C. Parmelee, R. D. Wade, S. Kumar, L. H. Ericsson, and K. A. Walsh, *Proc. Natl. Acad. Sci. USA 78*: 848 (1981).

79. S. Shoji, L. H. Ericsson, K. A. Walsh, E. H. Fischer, and K. Tetani, *Biochemistry 22*: 3702 (1983).

80. M. D. Uhler, J. C. Chrivia, and G. S. McKnight, *J. Biol. Chem. 261*: 15 (1986).

81. M. D. Uhler, D. F. Carmichael, D. C. Lee, J. C. Chrivia, E. G. Krebs, and G. S. McKnight, *Proc. Natl. Acad. Sci. USA 83*: 1300 (1986).

82. M. D. Uhler and G. S. McKnight, *J. Biol. Chem. 262*: 15 (1987).

83. K. Seguro and M. Motoki, *Agric. Biol. Chem. 53*: 3263 (1989).

84. F. F. Shih, in *Food Proteins* (K. D. Schwenke and R. Mothe, eds.), VCH, Weinheim, Germany, 1993, pp. 180–186.

85. N. F. Campbell, F. F. Shih, and W. E. Marshall, *J. Agric. Food Chem. 40*: 403 (1992).

86. L. F. Ross and D. Bhatnagar, *J. Agric. Food Chem. 37*: 841 (1989).

87. L. F. Ross, *J. Agric Food Chem. 37*: 1257 (1989).

88. L. A. Pinna, F. Meggio, and F. Merchiori, in *Peptides and Protein Phosphorylation* (B. C. Kemp, ed.), CRC Press, Boca Raton, FL, 1990, pp. 145–169.

89. J.-C. Mercier, F. Grosclaude, and B. R. Dumas, *Milchwissenschaft 27*: 402 (1972).

90. E. W. Bingham and H. M. Farrell, Jr., *J. Biol. Chem. 249*: 3647 (1974).

5
Preparation and Functional Properties of Protein–Polysaccharide Conjugates

Akio Kato

Yamaguchi University, Yamaguchi, Japan

I. INTRODUCTION

Many researchers have developed methods for improving the functional properties of proteins by using chemical [1–9] and enzymatic modifications [10–14] to meet the requirement for high-quality proteins in food ingredients. However, most of these methods are not used for food applications because of potential health hazards or the appearance of detrimental products. Therefore approaches different from the conventional ones are desirable for the improvement of the functional properties of proteins in food systems.

Proteins have unique surface properties due to their large molecular size and their amphiphilic properties. However, the industrial applications of food proteins are limited, because proteins are generally unstable against heating, organic solvents, and proteolytic attack. Therefore if proteins can be converted into stable forms, the applications of proteins in food processing will be greatly broadened. Glycosylation of proteins is expected to overcome their instability to heating and to improve their functional properties. Thus the glycosylation of proteins with monosaccharides or oligosaccharides has been attempted [7–9]. However, the improvements were not enough for industrial use. Marshall [15] reported that soluble protein–dextran conjugation by coupling proteins to cyanogen-bromide-activated dextran was dramatically effective in promoting the heat stability of various enzymes. We also found that the soluble protein–dextran conjugates prepared by coupling with cyanogen-bromide-activated dextran showed excellent emulsifying properties in addition to heat stability [16]. Based on these observations, we have developed a novel method of conjugating proteins with polysaccharides by spontaneous Maillard reaction in controlled dry heating between the ε-amino groups in proteins and the reducing-end carbonyl groups in polysaccharides; the resulting protein–polysaccharide conjugates showed excellent emulsifying properties superior to those of commercial emulsifiers [17–19]. Among the many chemical and enzymatic modifications of proteins to improve their functionality, this method could be one of the most promising for food applications, because of its safety and other advantages. In addition to providing a dramatic improvement in emulsifying properties, this approach improved the solubility of insoluble gluten [20], enhanced the antioxidant effect of ovalbumin [21], and broadened the bactericidal effect of lysozyme [22,23]. Therefore Maillard-type protein–polysaccharide conjugates can be used as proteinaceous food additives such as emulsifiers, antibacterial agents, and antioxidants. This chapter summarizes the properties of Maillard-type protein–polysaccharide conjugates from the viewpoint of the development of new types of food additives, medicines, and cosmetics.

II. PREPARATION AND BINDING MODE OF PROTEIN–POLYSACCHARIDE CONJUGATES

Maillard-type protein–polyaccharide conjugates can be efficiently prepared during storage of the freeze-dried powders of protein–polysaccharide mixtures (molar ratio 1:5) at 60°C for a given day under either 65% or 79% relative humidity in a desiccator containing saturated KI or KBr solution respectively in the bottom. The Maillard reaction between the ε-amino groups in protein and the reducing-end carbonyl group in polysaccharide is accelerated in the low water activity described above. The rate of reaction for the formation of the conjugates seems to depend on the conformation of the proteins. The casein–polysaccharide conjugate is formed within a day, while it takes about one or two weeks to form the lysozyme–polysaccharide conjugate. It seems likely that the rigid structure of the protein may suppress, and the unfolding structure accelerate, the formation of conjugates, because of the difference in the reactivity of the lysyl residues exposed outside between folded and unfolded proteins.

The binding mode was investigated using ovalbumin–dextran and lysozyme–dextran conjugates [18,24]. The SDS polyacrylamide gel electrophoretic patterns demonstrate single bands for protein and carbohydrate stains near the boundary between stacking and separating gels, indicating the formation of the conjugate between ovalbumin or lysozyme and dextran. The molecular weight and the decrease in free amino groups of protein–dextran conjugates suggest that about two moles of dextran are bound to ovalbumin or lysozyme respectively. Since the binding ratio is expressed as average values, it is probable that proteins bound with one to three moles of dextran may also exist in the protein–dextran conjugates. The analysis of the low-angle laser light scattering combined with HPLC suggests that these protein–polysaccharide conjugates can easily form oligomeric micelle structures in aqueous solutions due to the amphiphilic property [24]. The binding reaction (panel A) and mode (panel B) for the formation of protein–dextran conjugates are proposed as shown in Fig. 1. The peptide analysis of the lysozyme–dextran conjugate showed that an

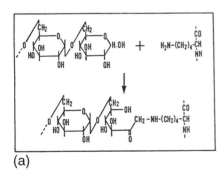
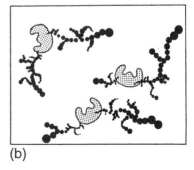

(a)   (b)

Figure 1 Scheme for the binding of protein with polysaccharide through Maillard reaction (a) and the binding mode (b). Dotted areas indicate protein molecules, and branched solid circles represent polysaccharide molecules. (From Ref. 25.)

active reducing-end group in dextran was attached to the ε-amino group in the lysine residues at positions 1 and 97 in lysozyme. The limited number of bound polysaccharides may come from the steric hindrance of the attached polysaccharide. This limitation is suitable for designing the functional properties of proteins, because the functions of proteins are deteriorated if most lysyl residues are masked by saccharides as observed in the conjugates of proteins with monosaccharides and oligosaccharides. This is the case for rigid and folded proteins. In contrast, unfolded proteins attach polysaccharides more than folded ones. Casein bound about four polysaccharides per mole in the conjugates [19], because the lysyl residues of unfolded protein are exposed outside and are easily reactive to the reducing-end carbonyl group in polysaccharide with a smaller steric hindrance than that of folded proteins.

III. EXCELLENT EMULSIFYING PROPERTIES OF PROTEIN–POLYSACCHARIDE CONJUGATES

Both protein and polysaccharide have a role in the stabilization of oil-in-water emulsions. Proteins adsorb at the oil–water interface during emulsification to form a coherent viscoelastic layer. On the other hand, polysaccharides confer colloid stability through their thickening and gelation behavior in the aqueous phase. Therefore protein–polysaccharide conjugates are expected to exhibit good emulsifying properties. As expected, the dramatic enhancements of emulsifying properties for lysozyme–polysaccharide, casein–polysaccharide, and soy protein–polysaccharide conjugates were observed (Fig. 2). The conjugates of proteins with polysaccharide revealed much better emulsifying activity and emulsion stability than control mixtures of proteins with polysaccharides. The emulsifying property of the conjugate of lysozyme with galactomannan was the best of various proteins. Similar excellent emulsifying properties were obtained in the conjugates of lysozyme with dextran. The use of galactomannan is desirable for food ingredients, because it is not so expensive as dextran and is already utilized as a thickener, binder, and stabilizing agent in food. In order to evaluate the potential to industrial applications, the emulsifying properties of dried egg white (DEW)–polysaccharide conju-

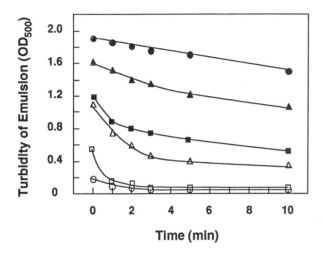

Figure 2 Comparison of emulsifying properties of various protein–polysaccharide conjugates. ●, lysozyme–galactomannan conjugate (dry heated for 2 weeks); ▲, α_{s1}-casein–galactomannan conjugate (dry heated for 1 day); ■, soy protein–galactomannan conjugate (dry heated for 1 week); ○, lysozyme–galactomannan mixture (no dry heating); △, α_{s1}-casein–galactomannan mixture (no dry heating); □, soy protein–galactomannan mixture (no dry heating).

gates were compared with commercial emulsifiers [25]. The DEW–galactomannan conjugates were much better than those of commercial emulsifiers (sucrose–fatty acid ester and glycerin–fatty acid ester). In addition, the emulsifying properties of the conjugates were not affected in acidic conditions, in the presence of 0.2 M NaCl, or by heating of the conjugates. Since high salt conditions, acidic pH, and/or heating processes are commonly encountered in industrial applications, the DEW–galactomannan conjugate may be a suitable ingredient for food processing. Since the commercial mannase hydrolysate (galactomannan) of guar gum is contaminated with considerable amounts of small molecular carbohydrates, thereby resulting in deterioration of emulsifying properties, the low-molecular-weight galactomannan should be removed prior to the preparation of the DEW–polysaccharide conjugate.

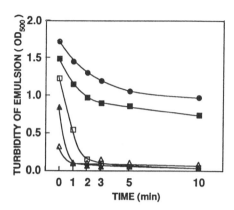

Figure 3 Effects of the length of saccharide chains on the emulsifying properties of lysozyme–saccharide conjugates. △, lysozyme; ▲, lysozyme–glucose conjugate; ☐, lysozyme–galactomannan (3.5–6.0 kDa) conjugate; ■, lysozyme–galactomannan (6.0–12 kDa) conjugate; ●, lysozyme–galactomannan (12–24 kDa) conjugate.

By screening various polysaccharides, galactomannan (mw 15,000–20,000) obtained from mannase hydrolysate of guar gum was found to be a suitable polysaccharide beside dextran. When glucose is attached to proteins in a similar manner, the functions of proteins are unfavorably lowered, and detrimental effects (browning color development etc.) are observed. As shown in Fig. 3, the emulsifying properties and heat stabilities of the conjugates are efficiently enhanced only when high-molecular-weight (more than 10 kDa) polysaccharides are used in the preparation of the conjugates [26]. Thus branched polysaccharides having a molecular size of more than 10 kDa are found to be efficient for the preparation of the conjugates.

In addition to the excellent emulsifying properties, improvements of various functional properties of Maillard-type protein–polysaccharide conjugates were reported from the viewpoint of the development of new types of food additives, medicines, and cosmetics. The examples reported so far are shown in Table 1. These are described in detail below.

Table 1 Improvement of Functional Properties of Protein–Polysaccharide
Conjugate Prepared by Maillard Reaction

Protein	Polysaccharide	Improved functional properties	Reference
Ovalbumin	Dextran	Emulsifying properties	*Agric. Biol. Chem. 54:* 107–112 (1990)
Lysozyme	Dextran	Emulsifying properties Antimicrobial action	*J. Agric. Food Chem. 39:* 647–650 (1991)
Gluten	Dextran	Emulsifying properties Solubility	*J. Agric. Food Chem. 39:* 1053–1056 (1991)
Casein	Dextran Galactomannan	Emulsifying properties	*Biosci. Biotech. Biochem. 56:* 567–571 (1992)
Lysozyme	Galactomannan	Emulsifying properties Antimicrobial action	*J. Agric. Food Chem. 40:* 735–739 (1992)
Ovalbumin	Dextran Galactomannan	Emulsifying properties Antioxidant effect	*J. Agric. Food Chem. 40:* 2033–2037 (1992)
Egg white	Galactomannan	Emulsifying properties	*J. Agric. Food Chem. 41:* 540–543 (1993)
Protamin	Galactomannan	Emulsifying properties	*J. Food Sci. 59:* 428–431 (1994)

IV. HEAT STABILITY OF PROTEIN–POLYSACCHARIDE CONJUGATES

As shown in Fig. 4, the heat stability of lysozyme was dramatically
increased by the conjugation with polysaccharide. No coagulation
was observed in the lysozyme–polysaccharide conjugates, while non-
glycosylated lysozyme formed insoluble aggregates during heating to
90°C with a transition temperature of 82.5°C. The lytic activity of
lysozyme in the conjugate was also recovered by cooling after heating to 90°C. These results suggest that the attachment of polysaccharide causes proteins to form stable structures. Upon heating in aque-

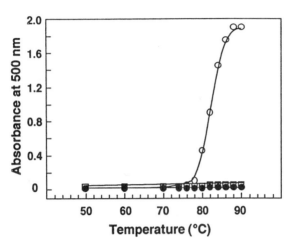

Figure 4 Thermal stability of lysozyme–polysaccharide conjugates. ○, lysozyme; ●, lysozyme–dextran conjugate; □, lysozyme–galactomannan conjugate. Lysozyme–polysaccharide conjugates were prepared by dry heating at 60°C for 2 weeks. Sample solutions (0.1%) in 67 mM phosphate buffer (pH 7.4) were heated at an increasing rate of 1°C/min. The resulting turbidity was measured with the absorbance at 500 nm.

ous solution, the protein molecule partially unfolds and results in aggregates due to the heat-induced disruption of the delicate balance of various noncovalent interactions. This process may be reversible in the protein–polysaccharide conjugates, because of the inhibition of the unfolded protein–protein interaction due to the attached polysaccharide. This resistance to heating is favorable for food applications, because heating is essential for the pasteurization of food ingredients. Therefore the approach of protein–polysaccharide conjugates can be useful for development of various functional food proteins.

V. NOVEL ANTIMICROBIAL ACTION OF LYSOZYME–POLYSACCHARIDE CONJUGATES

Many attempts have been made to develop food preservatives having superior antimicrobial effects without toxicities. For this purpose, hen egg white lysozyme may be one of the most promising ingredients. It is well known that lysozyme attacks only specific positions of glycosidic bonds between N-acetylhexosamines of the peptidoglycan layer in bacterial cell walls. However, since the cell envelope of these bacteria contains a significant amount of hydrophobic materials such as lipopolysaccharide (LPS) covered over the thin peptidoglycan layer, lysozyme fails to lyse gram-negative bacteria when it is simply added to the cell suspension in native form. As discussed in a previous paper [27], synergistic factors such as detergents and heat treatment destabilize and consequently solubilize the outer membranes, which are mainly composed of LPS. Therefore the excellent surfactant activity of lysozyme–polysaccharide conjugate seems to destroy the outer membrane of gram-negative bacterial cells synergistically along with thermal stresses.

From this viewpoint, the antimicrobial action of lysozyme–polysaccharide conjugates is expected. Despite the steric hindrance due to the attachment of dextran or galactomannan, the lytic activity of lysozyme was considerably preserved, and the antimicrobial action of lysozyme–dextran conjugate for typical gram-positive bacteria was almost the same as that of control lysozyme [22]. Antimicrobial effect of lysozyme–dextran conjugate on five different gram-negative bacteria was also observed (Table 2). The living cells were dramatically decreased with heating time at 50°C in the presence of lysozyme–dextran conjugate and completely disappeared from the medium after 40 min. On the contrary, the bactericidal effects were not observed in the presence of native lysozyme as well as in the control medium (buffer alone). The similar antimicrobial effects of lysozyme–galactomannan conjugate were observed for five gram-negative bacterial strains measured after heat treatment at 50°C for 30 min [23]. Although all strains tested were slightly affected by heating in the absence of the conjugate, the bactericidal effects on all strains were observed in the presence of lysozyme–galactomannan conjugate. Thus it was concluded that the lethal effect was effectively induced by exposing the cells to lysozyme–galactomannan conjugate as well as lysozyme–dextran conjugate.

Since lysozyme–polysaccharide conjugate is very stable to heat treatment and capable of perturbing the outer membrane of gram-

Table 2 Antimicrobial Activity of Lysozyme–Polysaccharide Conjugate

	Log survival ratio[a]			
Strains	Control[b]	LZ[c]	LDX[d]	LGM[e]
Gram-negative bacteria				
Aeromonas hydrophila	−3.95	*	*	*
Vibrio parahaemolyticus	−1.14	−2.84	*	*
Escherichia coli	−0.82	−0.79	*	*
Proteus milrabilis	−2.05	−2.15	*	*
Klebsiella pneumoniae	−0.59	−0.44	−2.21	−2.89
Gram-positive bacteria				
Bacillus cereus	−1.14	−1.85	−2.59	−2.61
Staphylococcus aureus	−1.87	−2.65	−2.70	−2.82

[a] Log survival ratio when tested strains were incubated at 50°C for 30 min. The values −1, −2, and −3 mean 1/10, 1/100, and 1/1000 survival ratios of cell numbers to the initial living cell numbers respectively.
[b] In control medium without lysozyme.
[c] In the medium supplemented with 0.05% native lysozyme.
[d] In the medium supplemented with 0.05% (for protein) lysozyme–dextran conjugate.
[e] In the medium supplemented with 0.05% (for protein) lysozyme–galactomannan conjugate.
*Nonsurvival (no living cells).

negative bacteria, it is good for food applications requiring heat treatments. In addition to the antimicrobial activity, lysozyme–galactomannan conjugate demonstrated better emulsifying properties than those of commercial emulsifiers under different conditions (in acidic pH or high salt solutions). Conjugates prepared without using chemicals can be applied in food formulas as safe multifunctional food additives. Some therapeutic effects of galactomannan may be expected. Yamamoto et al. [28] have reported that the oral administration of galactomannan decreases the total content of lipids in the livers of rats. Since galactomannan is not so expensive as dextran, lysozyme–galactomannan conjugate can probably be used as a food preservative as well as an emulsifier.

VI. IMPROVEMENT OF SOLUBILITY OF GLUTEN BY CONJUGATION WITH DEXTRAN

Insoluble gluten was solubilized by protease treatment, and then the high-molecular-weight fraction was conjugated with dextran [19]. The gluten–dextran conjugate thus obtained was soluble at a wide range of pH 2–12. In addition, the emulsifying properties of gluten–dextran conjugate were greatly increased. Thus protease digestion followed by polysaccharide conjugation is one of the most promising methods for utilizing insoluble protein sources. The solubilities of cereal and legume proteins are generally low. Therefore these proteins should be solubilized prior to conjugation with polysaccharides by using protease treatment and other methods. Deamidation may be another way to solubilize these proteins. We are trying to prepare protein–polysaccharide conjugates in combination with deamidation for these plant proteins.

VII. ENHANCEMENT OF ANTIOXIDANT EFFECTS OF OVALBUMIN BY CONJUGATING WITH POLYSACCHARIDE

Ovalbumin has a protective effect on lipid oxidation [29]. Although the effect is not so strong as with the synthetic antioxidants BHT or BHA, it can be used as one of the promising antioxidants. Thus an attempt was made to enhance the potential antioxidant effect by conjugating ovalbumin with polysaccharide [21]. The antioxidant action of ovalbumin was remarkably enhanced by the covalent binding of dextran or galactomannan. As expected, a significant improvement of the emulsifying property was also observed in the ovalbumin–polysaccharide conjugates. The resulting high affinity to oil suggests that the radical scavenging activity of ovalbumin may be elevated by covering oil surfaces with the conjugates. The development of a protein emulsifier having an antioxidant action may be useful for food applications because food emulsions are susceptible to rancidity through lipid oxidation.

Animal dose tests and bacterial mutagenesis tests were done to confirm the safety of protein–polysaccharide conjugates [21]. It has been shown that these conjugates are nontoxic for oral administration to mice and that they are negative for the Ames test and the rec assay. In addition, the effect of protein–polysaccharide conjugate on the proliferation of mammalian cells was also investigated to ensure the safety of the conjugates [26]. There was no inhibition of cell growth, suggesting no detrimental effects of the conjugates on the mammalian cell. Therefore these protein–polysaccharide conjugates are safe and may be fruitful products as novel macromolecular food ingredients.

VIII. POLYMANNOSYL LYSOZYME CONSTRUCTED BY GENETIC ENGINEERING

In order to evaluate the effect of glycosylation on the functional properties of protein on a molecular basis, we constructed two types of glycosylated lysozymes, a small oligomannose chain ($Man_{18}GlcNAc_2$)-linked form and a large polymannose ($Man_{310}GlcNAc_2$)-linked form using genetic engineering [30]. A large amount of polymannosyl lysozyme was predominantly expressed in the yeast carrying the lysozyme expression plasmid, while a small amount of oligomannosyl lysozyme was also secreted in the yeast medium. Complementary DNA encoding hen egg white lysozyme was subjected to site-directed mutagenesis to obtain the Asn-X-Thr/Ser sequence that is the signal for asparagine-linked glycosylation. At positions 49, 67, 70, and 103, the signal for N-linked glycosylation was created. Only the mutant lysozyme (G49N) whose glycine 49 was substituted with asparagine was expressed glycosylated lysozymes. The oligomannosyl and polymannosyl lysozymes were secreted in the yeast carrying cDNA of G49N mutant. The polymannosyl lysozyme showed remarkable heat stability and excellent emulsifying property [31] (Fig. 5). Interestingly, the thermal stability and emulsifying property of polymannosyl lysozyme were much higher than those of oligomannosyl lysozyme. This result suggests that the attachment of polysaccharide is much more effective for the improvements of thermal stability and emulsifying property of proteins than that of oligosaccharide. It is probable that polysaccharide attached to lysozyme may stabilize the

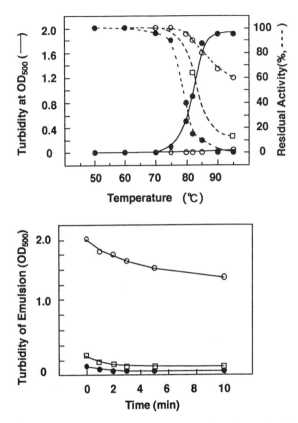

Figure 5 Thermal denaturation curves (upper) and emulsifying properties (bottom) of polymannosyl and oligomannosyl lysozymes constructed by genetic engineering. ●, wild-type lysozyme; ○, polymannosyl lysozyme; □, oligomannosyl lysozyme.

aqueous phase around the protein molecule, thereby causing the stabilization of the protein structure against heating. Similarly, the role of polysaccharide in the stabilization of emulsion is considered as follows. The polysaccharide orients to the aqueous layer around the oil droplets after the emulsion preparation and accelerates the formation of a thick steric stabilizing adsorbed layer around the emulsion, thereby inhibiting the coalescence of oil droplets. Thus it was shown on a molecular basis that dramatic improvements of the thermal stability and emulsifying property of lysozyme are brought about by the attachment of polysaccharide but not of oligosaccharide.

IX. CONCLUSIONS

Novel and promising approaches were described here to improve the functional properties of food proteins. The conjugation of proteins with polysaccharides through the Maillard reaction in a dry state was very effective in improving the thermal stability and emulsifying property of proteins. In addition, the lysozyme–polysaccharide conjugate showed novel antimicrobial action against gram-negative bacteria, and the ovalbumin–polysaccharide conjugate exhibited enhanced antioxidant action. The importance of the size of the saccharide chain in the effective improvement of the functional properties was proved by the genetic modification of lysozyme.

REFERENCES

1. Z. Haque and M. Kito, *J. Agric. Food Chem. 24*: 504–510 (1976).
2. C. H. Wu, S. Nakai, and W. P. Powrie, *J. Agric. Food Chem. 31*: 1225–1230 (1983).
3. N. Matsudomi, T. Sasaki, A. Kato, and K. Kobayashi, *Agric. Biol. Chem. 49*: 1251–1256 (1985).
4. J. E. Kinsella, *CRC Crit. Rev. Food Sci. Nutr. 7*: 219–280 (1976).
5. L. C. Sen, H. S. Lee, R. E. Feeney, and J. R. Whitaker, *J. Agric. Food. Chem. 29*: 348–354 (1981).
6. G. Matheis, M. H. Penner, R. E. Feeney, and J. R. Whitaker, *J. Agric. Food. Chem. 31*, 379–387 (1983).
7. J. W. Marsh, J. Denis, and J. C. Wriston, Jr., *J. Biol. Chem. 252*: 7678–7684 (1977).
8. M. J. Krantz, N. A. Holtzman, C. P. Stowell, and Y. C. Lee, *Biochemistry 15*: 3963–3968 (1976).
9. N. Kitabatake, J. L. Cuq, and J. C. Cheftel, *J. Agric. Food. Chem. 33*: 125–130 (1985).
10. M. Watanabe, A. Shimada, E. Yazawa, T. Kato, and S. Arai, *J. Food Sci. 46*: 1738–1740 (1981).
11. M. Watanabe, H. Toyokawa, A. Shimada, and S. Arai, *J. Food Sci. 46*: 1467–1469 (1981).

12. A. Kato, A. Tanaka, Y. Lee, N. Matsudomi, and K. Kobayashi, *J. Agric. Food. Chem. 35*: 285–288 (1987).
13. N. Nio, M. Motoki, and K. Takinami, *Agric. Biol. Chem. 49*: 2283–2286 (1985).
14. M. Motoki, K. Seguro, N. Nio, and K. Takinami, *Agric. Biol. Chem. 50*: 3025–3030 (1986).
15. J. J. Marshall and M. L. Rabinowitz, *J. Biol. Chem. 251*: 1081–1087 (1976).
16. A. Kato, K. Murata, and K. Kobayashi, *J. Agric. Food Chem. 36*: 421–425 (1988).
17. A. Kato, Y. Sasaki, R. Furuta, and K. Kobayashi, *Agric. Biol. Chem. 54*: 107–112 (1990).
18. A. Kato and K. Kobayashi, in *Microemulsions and Emulsions in Foods* (M. El-Nokaly and D. Cornell, eds.), ACS Symposium Series *448*, 1991, pp. 213–229.
19. A. Kato, R. Mifuru, N. Matsudomi, and K. Kobayashi, *Biosci. Biotech. Biochem. 56*: 567–571 (1992).
20. A. Kato, K. Shimokawa, and K. Kobayashi, *J. Agric. Food Chem. 39*: 1053–1056 (1991).
21. S. Nakamura, A. Kato, and K. Kunihiko, *J. Agric. Food Chem. 40*: 2033–2037 (1992).
22. S. Nakamura, A. Kato, and K. Kobayashi, *J. Agric. Food Chem. 39*: 647–650 (1991).
23. S. Nakamura, A. Kato, and K. Kobayashi, *J. Agric. Food Chem. 40*: 735–739 (1992).
24. A. Kato, K. Kameyama, and T. Takagi, *Biochim. Biophys. Acta 1159*: 22–28 (1992).
25. A. Kato, K. Minaki, and K. Kobayashi, *J. Agric. Food Chem. 41*: 540–543 (1993).
26. A. Kato, *J. Jpn. Soc. Food Sci. Tech. 41*: 304–310 (1994).
27. K. Nakamura and S. Mizushima, *Biochim. Biophys. Acta 413*: 371–393 (1975).
28. T. Yamamoto, S. Yamamoto, I. Miyahara, Y. Matsumura, A. Hirata, and M. Kim, *Denpun Kagaku 37*: 99–105 (1990).
29. M. Goto and K. Shibasaki, *J. Jpn Soc. Food Sci. 37*: 277–283 (1971).
30. S. Nakamura, H. Takasaki, K. Kobayashi, and A. Kato, *J. Biol. Chem. 268*: 12706–12712 (1993).
31. S. Nakamura, K. Kobayashi, and A. Kato, *FEBS Letters 328*: 259–262 (1993).

6
Enzymatic Modification as a Tool for Alteration of Safety and Quality of Food Proteins

Gyöngyi Hajós

Central Food Research Institute, Budapest, Hungary

I. INTRODUCTION

The basic function of food proteins is to keep us alive and healthy. The importance of the quality of proteins for food and medical uses was recognized many years ago. However, the meaning of protein quality [1] has been a subject of great polemic ever since. Much remains to be learned for the understanding of the role of proteins in food. Knowledge of the influence of structure on functional properties and biological activities of proteins is essential for their rational use in foods and for the improvement of the safety of food proteins.

The modification of proteins for human use as the origin of biochemistry has ancient roots. Among biochemical or biotechnological procedures, proteolysis and proteolytic modification are of special importance from the point of view of optimizing food proteins with respect to their nutritional and functional quality and biological activity. The goal of enzymatic modification is to formulate nutritious, economic foods mainly of plant origin that fit into healthy lifestyles using high-quality proteins. However, plant proteins are usually deficient in one or more of the essential amino acids.

This text was revised by Lilly Vámos-Vigyázó, D.Sc.

This chapter deals with protein structure–function relationships resulting from proteolytic modification, covalent amino acid incorporation, and the effects of these enzymatic modifications on sensory and nutritional quality, on the functional properties, and on the biological activities of proteins and peptides.

II. PROTEIN STRUCTURE, SURFACE, FUNCTIONS, AND APPROACHES TO THEIR MODIFICATION

Proteins are really marvelous molecules. The rigorously well-defined arrangement of a *protein molecule* is an organic order [2]. There is a cooperative interaction among the amino acids, more exactly among the great number of electrons and nuclei. The special properties, the biological functions of the proteins are due to this superior organization of the molecule. The amino acids commonly found in proteins can combine to form an almost endless number of different primary sequences. The versatility, the possibility, and the variability of the protein structures are enormous.

This particularly organized structure determines a well-defined surface of the protein. Recently, special attention has been focused on the *surface activity of food proteins* as the most important factor related to the functional properties and biological character of the proteins. The hydrophobic regions of the globular proteins are in general in the interior of the molecule, and the hydrophilic regions constitute its surface. Thus the globular proteins are less surface active in their native form.

Enzymatic modification reactions are mostly suitable for inducing conformational changes. These changes in conformation may result in the forming of highly surface-active proteins during particular reaction conditions. Covalent attachment of a highly hydrophobic amino acid ester to a sufficiently hydrophilic protein results in a protein with an amphiphilic surface. The flexibility of a globular protein is very important from the aspect of the functional properties, especially at the interface of the molecule. Hydrophobic, electrostatic, and steric parameters are significant in describing the protein molecules. The steric parameters, the *conformation* of the proteins, correlate better to the functionality than the variety of their sequence [3].

In the past ten years, protein modification has been the subject of many excellent publications and reviews [4,5,6,7,59]. Chemical or enzymatic food protein modification is theoretically feasible [8,9,10,175]. *Modification of proteins* may alter their net charge, hydrophobicity, structure, and furthermore their surface activity and thus the functional properties of the proteins.[11,12].

Chemical protein modification is not very desirable for food applications because of the drastic reaction conditions, the nonspecific chemical reagents, and the difficulty of removing residual reagents from the final product.

Enzymatic modification provides several advantages including fast reaction rates and mild conditions. This will result in reduced energy costs and increased processing efficiency. Enzymatic methods are more attractive, because most enzymes can be produced in large quantities, each having appropriate physical, chemical, and catalytic properties [7]. However, the most important feature of enzymatic methods is the specificity of the enzymes. A computer program would be very useful in predicting the possible desired changes in protein structure owing to a chemical or an enzymatic modification. Programs may [13] predict what differences in the sequence of a protein cause the structural difference. The designed chemical and/or enzymatic modification may result in alteration in the surface structure of the proteins and enzymes and might induce improvement in their functional properties.

Modification of proteins by transglutaminase [14,15,16,17,18], peptidoglutaminase [19,20,21,22], and protein kinase [23,24] on a laboratory scale has been reported. Whitaker [25] discussed in great detail the impact of these potential modifications on the structure and properties of the proteins. However, the effect of *proteinases* on proteins has been extensively investigated, and proteinases are the only protein-modifying enzymes currently in commercial use. However, immobilizing proteases proved to be very effective in enzymatic peptide modification [26]. From the aspect of human health and safety, the use of proteinases in protein modification should offer better products than those obtained by corresponding chemical modification methods. However, proteolysis of a protein depends not only on a function of the enzyme and the structure of the protein but also on the environmental conditions of the reaction. When a protein is transformed from a globular form into a looser structure, its surface activity is altered. The enzymatic modification of proteins results in changing their conformation and consequently their functional, physicochemical, and biological properties.

III. HYDROLYSIS AND PEPTIDE BOND SYNTHESIS BY PROTEINASES

One of the basic questions of food science is how to improve the quality of food proteins, i.e., how to modify their structures, their surface activities. The enzymatic modification of proteins might have a good chance of becoming an attractive means of altering the structure and improving the nutritional, sensory, and functional quality of foods. Enzymatic reactions have recently been developed in two directions: (1) the application of enzymes under particular circumstances (special physical and chemical conditions) and (2) the development of enzymes of particular specificity for special purposes. Food proteins have been modified (a) by using *endopeptidases* for producing fragments from peptides and/or proteins with properties different from the original substrates and (b) by using *exopeptidases* for changing the terminal amino acids of the peptide chains in a desired manner.

The ability of proteinases to catalyze peptide bond hydrolysis and synthesis has long been known [29,30,31,32,33,34], and there is a renewed interest in it in present food protein biochemistry [10,25,27,28,35,36,37,59].

The *enzyme-catalyzed hydrolysis* of a particular peptide bond is determined by two major factors, the susceptibility of that bond to the specific proteinase and the flexibility of the protein chain in the region of the bond. Thus a globular protein undergoes frequent fluctuations, and the sites of the peptide chains of highest mobility are the most susceptible to proteolytic reactions. On the other site, the flexibility of proteins may be detected by their susceptibility to proteases [38]. This susceptibility is also suitable for detecting the relationship between functional surface properties and the flexibility of proteins. The enzymatic hydrolysis of food proteins [39] produces peptides of smaller molecular sizes and less secondary structure than that of proteins.

The direction of an enzymatic equilibrium reaction (hydrolysis or peptide bond synthesis) depends on the chemical conditions [40]. Besides the above-mentioned hydrolysis, the proteinase-catalyzed synthesis of peptide bonds [41] has also been widely investigated and will be discussed below.

The history of reverse proteolysis, the enzymatic modification or the so-called plastein reaction, has had some very interesting periods. At the beginning of our century several authors showed [29] that proteineous products were forming in proteolytic hydrolysates of proteins in the presence of high peptide concentrations and of an appropriate proteinase. Later, during the 1940s, many biochemists believed that the plastein reaction might be involved in in vivo protein biosynthesis [42,43,44]. In the 1960s, when the theory of peptide chain elongation in vivo was proved, the plastein reaction was no more in the focus of interest. However, in the 1970s, a Tokyo research group [45,46,47,48,49] reinvestigated the plastein reaction with the object of applying it to the improvement of food proteins. The conditions required for the *enzymatic peptide bond synthesis and/or transpeptidation* are as follows.

A. Required Conditions

First: the *substrate* must be an enzymatic protein hydrolysate (consisting of a mixture of low-molecular-weight peptides) with an average molecular weight less than 1000 [50] between 450–1450 [51], with soy proteins: 1043 and 635 [52].

Second: the *substrate concentration* should be 15–40% w/v [53]. A lower substrate concentration is more favorable for the direction of hydrolysis [86].

Third: the *pH* is of special interest during the enzymatic reaction [49]. The optimum pH value for hydrolysis differs from that for the synthesis with serine proteinases and aspartic proteinases. However, the optimum pH is the same for both protein hydrolysis and peptide bond synthesis [49] in the presence of cysteine proteinase as catalyst. Different theories have been proposed to explain the mechanism of these enzyme catalyzed reactions.

B. Main Forces of the Proteolytic Reactions

Condensation is believed by some authors to be the main force of this enzymatic reaction [43,44,54,55]. Tsai and coworkers [52] found a significant increase in the molecular mass of the enzymatically modified protein as a result of polycondensation reactions.

Noncovalent forces have also been reported to play an important role in these enzyme-catalyzed reactions [50,56]. Moreover, Hofsten and Lalasidis [50] were of the opinion that covalent forces did not play a role in these reactions. Several investigators have shown that the product produced during the plastein reaction is composed of aggregates held together by hydrophobic and ionic bonds [57,58]. Others [59] emphasized an entropy-driven aggregation process. *Transpeptidation* has been considered by a number of authors [46,60,61] as the mechanism of enzymatic modification processes (resynthesis, plastein reaction, EPM). That means that a great number of peptide bonds are split and new covalent bonds formed in the course of the enzymatic process.

During *hydrolysis* catalyzed by serine proteases an acyl–enzyme complex transfers the acyl group to water. However, in *enzymatic synthesis*, the acyl group is not transferred to water but to a nucleophile, that is, to an amino group or amino acids/peptides. Thus a transpeptidation reaction takes place during the enzymatic modification [46,57].

Lozano and Combes [65,69] proposed an overall interpretation of all proteolytic reactions: hydrolysis, transpeptidation, and condensation, as a general mechanism of the plastein reaction.

An *enzymatic peptide modification* (*EPM*) process has been recently elaborated for the tailoring of peptides and proteins [66,82,123,135]. The goal of this method was production of the following products of special nutritional character:

Methionine enriched products, in order to increase the content of the methylating agent in the modified protein chains
Special amino acid (methionine) enriched peptides, for improvement of the biological value
Proteins and peptides of reduced allergenic character

The molecular mass characteristics of the product produced in the enzymatic modification reaction are of considerable interest. Polyacrylamide gel electrophoresis in the presence of sodium dodecyl sulfate was used to detect changes in molecular mass during the enzymatic resynthesis. These investigations were carried out from casein hydrolysate produced with pronase, α-chymotrypsin, or papain [62]. The electrophoretic patterns obtained from the different enzymatically modified products showed remarkable changes in molecular mass distribution, which could be mainly ascribed to transpeptidation reactions [63]. The presence of high amounts of amino acid derivatives in the reaction mixture [64] significantly modified the process of transpeptidation. Hydrolysis and synthesis of peptide bonds as consequences of EPM [66,67] have been studied by determining the degree of hydrolysis (DH) using the TNBS method. Correlation between DH and reaction time could be described by power curves. For EPM without amino acid addition, a moderately increasing curve, while for EPM with Met addition, a slightly decreasing power curve was found to fit in satisfactorily with the experimental values [64]. Since in these experiments, in the presence of SDS, which strongly promotes dissociation, a decrease in DH has always been observed, the formation of covalent bonds is supposed to be highly probable. This is in agreement with our earlier findings.

The determinations of the degree of hydrolysis in the course of the enzymatic reaction reveal that transpeptidation is the major process in enzymatic peptide modification. The structural changes in EPM products produced from casein were also investigated by isoelectric focusing. For the evaluation of the electrophoretograms, a computer-assisted method has been elaborated. The results corroborated that transpeptidation took place during EPM treatment [68]. The separation, by the net charge of protein fractions, of the EPM products with and without Met enrichment was found to be significantly different from that of the substrate. Since the average molecular mass remained practically unchanged during the EPM process [63], these differences indicated that transpeptidation took place in the reaction.

On the basis of the large number of nonhomologous bands in both cases, we concluded that a great number of peptide bonds was cleaved and formed due to considerable transpeptidation. It is interesting to note that, in this respect, the samples both with and without methionine addition behaved in an analogous way. The methionine content of the EPM product with Met enrichment was higher by 2.6% only than that of the product without amino acid enrichment. Nevertheless, the peptide fractions of the EPM product without amino acid incorporation and the Met-enriched product differed from each other remarkably. This is why we suppose that the presence of methionine in the reaction mixture and its incorporation give rise to a new type of transpeptidation, namely, they modify the sequential position of the cleaved and formed peptide bonds.

C. Hydrophobic Character of the Reaction Products

There is still much debate in the literature about the actual reaction mechanism [54,56,57,71] and thereby also about the hydrophobic character of the reaction products of the enzymatic peptide modification. Noar and Shipe [56] concluded that the incorporation of methionine by one-step enzymatic process [72,73] was due to the formation of hydrophobic bonds.

The plastein reaction was also investigated for the possible use of proteins in novel food applications. This was done via establishing a relationship between the microenvironmental conditions of the plastein reaction and the amino acid composition of the products. At low substrate concentration, produced by manipulation with additives, the plastein reaction was enhanced via the condensation pathway [74,75]. The plastein activity in the presence of α-chymotrypsin as catalyst increased with substrate concentration in the range of 10–30% (w/v). The substrate of this plastein reaction was a peptic hydrolysate of albumin obtained at 40°C, pH 7.0. The content of hydrophobic amino acids, in this case Ile, Leu, Val, and Tyr, increased in the plastein products, while the content of Asp, Glu, Ser, and

Thr decreased. Interestingly, the increase in substrate concentration resulted in basically different amino acid compositions of the plastein products.

Andrews [76] reported that plasteins were formed by an association of predominantly hydrophobic peptides via hydrophobic and possibly ionic bonding. Aso and coworkers [77,78] concluded that hydrophobic forces were a major factor in plastein chain assembly. They found that, compared with the substrate, the water-insoluble product contained smaller ratios of hydrophilic and larger ratios of hydrophobic amino acid residues. The results of Sukan and Andrews [58] showed that hydrophobic amino acids such as phenylalanine, leucine, isoleucine, tyrosine, valine, and proline were preferentially incorporated into plastein at the expense of hydrophilic amino acids. Also others have reported on a trend of preferential incorporation of hydrophobic amino acids into the protein product in enzyme-catalyzed reactions [46,60,79,80,81].

Although certain hydrophobic amino acids were found to concentrate in some EPM samples in exceptional cases, investigation into this respect carried out with soy protein and casein did not provide any evidence for generalizing this finding [82]. However, in the case of enzymatic modification of an enzymatically prehydrolyzed milk protein concentrate, the analysis of the product showed a significantly higher Met incorporation into peptides containing a relatively high ratio of apolar amino acids [83].

D. Effects of Additives in the Reaction Media on the Enzymatic Modification Process

Effects of various naturally occurring nonprotein substances (carbohydrates, polysaccharides, fats, and salts) on enzymatic hydrolysis of soy protein isolate and plastein formation from hydrolyzed soy protein were investigated by Hagan and Villota [70]. Several nonprotein substances were found to interfere with the protein solubility in 10% trichloroacetic acid assay. Positive interference was noted for systems containing saturated and unsaturated fatty acids and magnesium, but negative interference was observed for systems containing guar gum, xanthan gum, calcium chloride, and gum arabic.

The amino acid compositions of the plastein products obtained in the presence of polyols in the reaction mixture were very similar. The results indicated that the effect of reduction in water activity induced by polyols was similar to that of increased substrate concentration [84]. The activity of the plastein reaction was investigated as a function of the concentration of different hydroxylated additives using a peptic hydrolysate of BSA. The catalyst of the plastein reaction was α-chymotrypsin. The authors found that the presence of polyols in the plastein reaction resulted in products in the following order: hydrolysis < transpeptidation < condensation, that is, it increases the role of condensation in the plastein reaction. A final conclusion of this work is that in the α-chymotrypsin–catalyzed plastein reaction, all possible biocatalytic pathways can occur simultaneously: hydrolysis, transpeptidation, and condensation. This work shows a new way for the enzyme-driven condensation of specific peptides in aqueous media at relative low (10% w/v) substrate concentration.

The presence of salts in the enzyme-catalyzed reactions favored plastein production proportionally to their concentration. However, the conformation of the protein was changed by the influence of salts [75]. Léonil et al. [85] reported that the pH effect on the precipitation of peptides is more marked than the effect of salt. Investigations were carried out into the precipitation of a group of hydrophobic peptides arising from a tryptic casein hydrolysate.

E. Enzymatic Reactions in Organic Media

Proteolytic reactions take place in general in aqueous media, with the equilibrium in the direction of hydrolysis [87]. However, with decreased water activity, the equilibrium of the reaction is shifted toward synthetic reactions [74]. The use of organic media provides a potentially useful approach to the enzymatic modification of food protein. It has been already established that proteinases are active in organic media. They catalyze ester synthesis or aminolysis by synthesizing new peptide bonds preferentially to hydrolysis.

Polyethylene glycol is used to make the enzymes soluble in organic solvents [88], and high reaction yields have been obtained with polyethylene-glycol–modified chymotrypsin [89], and papain in benzene [90]. Enzymatic modification reactions with deacylation, via aminolysis, of an intermediate covalent acyl-enzyme also support the mechanism of transpeptidation in kinetically controlled peptide synthesis in organic solvent [91]. Side reactions in enzymatic peptide synthesis in organic media were also investigated together with the effects of enzyme, solvent, and substrate concentrations [92]. According to the results, with activated esters as acyl-group donors, the enzymatic peptide synthesis can be considered kinetically controlled both in low-water systems and in water.

For peptide synthesis reactions carried out in the presence of methanol, a partial deactivation of α-chymotrypsin was also reported [93].

Enzymic catalysis in organic media promotes aminolysis over hydrolysis [94]. The role of condensation increases with the presence of water-activity reducing additives during the plastein reaction [65].

IV. COVALENT AMINO ACID ENRICHMENT OF PEPTIDE CHAINS

During enzymatic modification under appropriate reaction conditions, L-amino acids (generally in ester form) are partially covalently incorporated into the peptide chains of a protein hydrolysate. Thus these enzymatic modification reactions with amino acid enrichment would be expected to be more important for health aspects than other modification processes without covalent amino acid enrichment.

Aso et al. [95] studied a model system in order to obtain basic information on the mechanism of amino acid incorporation during an enzymatic modification reaction in the presence of papain. They found that the amino acid ester reacted as a nucleophile in the aminolysis of the acyl-enzyme intermediate to result in the formation of new peptides. Several proteases used in enzymatic peptide bond synthesis are known to form transitory acyl-enzyme intermediates during the hydrolysis of proteins. However, the acyl groups can be transferred to other nucleophiles (amino terminals of peptides or amino acids), synthesizing new peptide bonds [71]. With full knowledge of the above-mentioned facts, covalent amino acid enrichment of proteins can result in

Improving their nutritional value
Modifying their functional properties
Altering their surface activity and their functional properties and biological activity

The first procedure using enzyme catalysis for introducing appropriate essential amino acids into food proteins—the "plastein reaction"—was proposed by Yamashita et al. [49,96] and Fujimaki et al. [46]. Several groups from many countries have also investigated the enzymatic modification reaction of proteins [97]. The authors mainly reported on a good yield when incorporating amino acids (methionine) into proteins (soy protein) [72,73,90,98,99]. Effective methionine enrichment of soy proteins was reported also in the presence of papain as catalyst [100]. Interestingly, n-hexyl esters of several amino acids were much more reactive than their ethyl esters [101].

Aso et al. [102] found as a result of enzymatic modification that the covalently attached methionine was localized at or near the C-terminals of the protein chains, probably in the form of oligomers. Protease-induced oligomerization of α-amino acid esters was also observed [103].

Methionine-enriched protein was produced also from an enzymatically prehydrolyzed milk protein using an enzymatic peptide modification method with α-chymotrypsin as catalyst. Amino acid incorporation leading to methionine enrichment of the product proceeded via formation of covalent bonds. The concentration of the substrate was 25% (w/v). Methionine was added to the reaction mixture in the form of methionine methyl ester hydrochloride. An ester/substrate ratio of 1:5 was used in the enzymatic peptide modification reaction. The methionine content of the product was twice as high as that of the substrate. The slight change in the degree of hydrolysis revealed that part of the amino acids were bound to the peptide chains and that transpeptidation was the main force during this enzyme-catalyzed reaction. The newly incorporated Met was located in C- and N-terminals in a ratio of 3:1 [82].

The location of methionine incorporation into peptide chains by enzymatic modification was also investigated using L-methionine-S-methyl[14]C methyl ester hydrochloride and L-[3H]methionine ethyl ester hydrochloride [104]. Substrates of the enzymatic modification used were a tryptic hydrolysate of serum albumin and an α-chymotryptic hydrolysate of casein. α-chymotrypsin was used as catalyst during the EPM reactions. Part of the L-methionine-S-methyl[14]C methyl ester was incorporated as Met into peptide chains. A maximum curve was found to describe the relation between α-chymotrypsin–induced covalent incorporation of Met and the ratio of L-^{3H} methionine ethyl ester to protein hydrolysate. These ensure great versatility to the enzymatic peptide modification process.

The covalent incorporation of Met into the α-chymotryptic hydrolysate of buffalo's milk proteins in the presence of α-chymotrypsin as catalyst was recently investigated [123]. Both the covalent Met incorporation and the methionine concentration in the reaction mixture can be seen in Fig. 1. The proteinase-induced covalent incorpo-

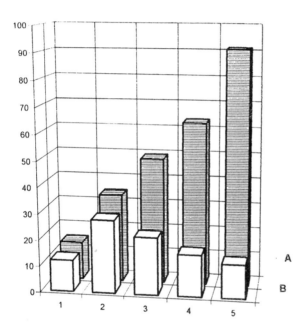

Figure 1 Methionine contents ▤ added to the reaction mixtures and ▢ covalently bound in the EPM products. ▤ The methionine contents in the different reaction mixtures are as follows: (1) 0.15 g Met/1g hydrolysate; (2) 0.34 g Met/1g hydrolysate; (3) 0.48 g Met/1g hydrolysate; (4) 0.63 g Met/1g hydrolysate; (5) 0.92 g Met/1g hydrolysate. ▢ Content of the covalently incorporated methionine in the EPM products. The methionine contents (percentage of the total amino acid content) were determined by amino acid analysis. A, Met concentration in the reaction mixtures (g/100 g); B, Met enrichment (%).

ration of methionine shows an optimum curve in the function of the methionine concentration in the reaction mixture. The optimum of the α-chymotrypsin–induced Met enrichment was at the ratio of 0.34 g Met/1 g hydrolysate in this reaction. These results suggest that the change in the ratio of the amino acid ester and the peptides of the proteolysate alters the mode of the transpeptidation and the effectiveness of the amino acid enrichment.

Methionine-enriched protein was produced also from an enzymatically prehydrolyzed milk protein (SR) by the EPM process, and the methionine contents of SR and the EPM product were determined by a microbiological method. The methionine content in the EPM product was more than twice as high as in the substrate. These protein fractions were separated by thin-layer chromatography [83]. The separated peptides and peptide mixtures were eluted and their molar amino acid contents were determined. The ratio of polar and apolar amino acids in the peptides was found to be influenced basically by transpeptidation taking place in the EPM reaction. The analysis of the peptides of these products showed that methionine was incorporated mainly in peptides with a relatively high ratio of apolar amino acids.

Recent studies also supported covalent amino acid incorporation during the enzyme-catalyzed modification reaction [105,106], and proteolysis in organic solvents is also mentioned as a particular way of amino acid incorporation involved in aminolysis [107].

V. EFFECTS OF ENZYMATIC MODIFICATION OF PROTEINS AND PEPTIDES

A proteinase-catalyzed reaction including splitting and synthesis of peptide bonds is a process also suitable for covalent amino acid incorporation into peptide chains. This type of enzymatic modification reaction of food proteins is useful for different purposes: alteration of sensory properties, solubility, nutritional quality, functional properties, antifreeze character, and different biological activities. Recently, special proteinase-catalyzed reactions have been elaborated by which proteins can be modified with particular respect to their primary structure and conformation.

A. Alteration of Sensory Properties

A Tokyo group [46] was the first to propose a combined process of enzymatic protein hydrolysis and resynthesis for producing a product with improved sensory properties and modified amino acid composition. An enzymatic reaction was used also for the removal of bound impurities [108,109], for debittering of hydrolysates [47,110], and for decolorization of proteins of particular origin [111].

Bitterness can be ascribed to the formation of peptides with high content of hydrophobic amino acids [112]. Protein hydrolysates are produced for food and special therapeutic uses because of their high absorbability and good nutritional quality. Therefore controlling enzymatic reactions is important for obtaining hydrolysates of little bitterness. As the existence of bitter peptides appears to be the main disadvantage in using partially hydrolyzed casein for food purposes. Tchorbanov and Iliev [113] recommended high ethanol concentration in the reaction medium for proteolytic processes. Under these circumstances, trypsin and α-chymotrypsin action led to products with reduced bitterness. Casein hydrolysates without any bitter taste can be produced in the presence of 40–60% (w/v) ethanol in the reaction mixture using a proteinase with narrow specificity.

Peptic hydrolysis was also applied [114] for food use with chicken heads. The final dried products were pale cream colored, had no bitter taste, were of high microbiological quality, and had a high mineral content. These highly soluble peptic hydrolysates would appear to be useful as flavoring and seasoning additives.

B. Changes in Solubility

Solubility is due to the native structure of proteins. In soluble proteins, hydrophilic parts are located on the outside of the molecule, and the hydrophobic regions are buried. Denaturation induces the insolubility of a protein. According to Bigelow [115], two structural features, hydrophobicity and charge frequency, are the major factors that control the solubility of proteins.

The ratio of the water-insoluble/soluble protein fractions may change during the enzyme-catalyzed reaction [78]. Insolubility of proteins might be altered also by incorporating certain amounts of hydrophilic amino acids into enzymatic hydrolysates of the proteins. α,γ-diethyl ester of glutamic acid was incorporated into the peptide chain during a plastein reaction, producing a modified soy protein with higher solubility [108,116]. Glutamic acid diethyl ester was incorporated also into a hydrolysate of fish proteins. The product showed higher water-holding capacities and increased water dispersibilities. Proteolysis is one of the most effective methods for improving the solubility of proteins [117]. Enzymatic modification is an effective way of increasing solubility and preventing heat coagulation also with soy proteins [118].

C. Improvement of Nutritional Quality

The biological value of food proteins is greatly influenced by the ratio of their essential amino acids. Most food proteins, first of all plant proteins, are deficient in some essential amino acids. The limiting essential amino acid in legume and milk proteins is methionine, and in cereal proteins it is lysine. Beside the knowledge of the requirements of quality and composition of amino acids, it is equally important to know whether these amino acids can be best utilized as free amino acids, in the form of peptides or proteins. The following possibilities exist for improving the balance of essential amino acids in the proteins:

1. Mixing the protein with another one containing complementary essential amino acids
2. Fortification of the protein with the limiting free amino acids or with polyamino acids
3. Modification via genetic engineering
4. Modification via covalent enrichment of proteins/peptides with the limiting essential amino acids

   1.   Choosing proteins with different amino acid contents for complementation by mixing in an appropriate ratio is a very difficult and not always available method. One of the main problems is, for instance, that methionine usually is one of the limiting essential amino acids in all proteins of nonanimal origin.
   2.   The fortification of a protein with the appropriate free essential amino acid(s) is not always the best process for food products. The disadvantages of this method are (i) the use of free amino acids

for food purposes is not recommended above a critical level for toxicity reasons, (ii) the loss of added amino acids during food processing, (iii) the utilization of free amino acids is not always desirable because of the competition during absorption (at the same time and in high concentration). According to the current understanding of the science of nutrition, the biological utilization of mixtures of free amino acids is much less effective than that of foods consisting of peptides and proteins of the same composition [119,120,121]. In special cases, peptide mixtures obtained by partial enzymatic cleavage of proteins are more advantageous than unmodified proteins, for instance in dietary food or when a large quantity is required within a short time. This frequently occurs with athletes or in special diet problems (e.g., food allergy). The digestion and absorption of dipeptides and tripeptides is more effective than that of an equivalent amount of the appropriate free amino acids [122].

The results of animal feeding experiments equally confirmed the statements of the literature that supplementation with covalently bound amino acids results in a significant improvement of the in vivo biological value. Although free amino acid supplements often increase the biological value, the utilization of these products is not satisfactory, since living organisms need amino acids bound in peptide bonds. During protein digestion in the human gut, the concentration of amino acids in the intestinal lumen increases in the form of peptides more than in the form of free amino acids. This observation has established that for intestinal absorption peptides are quantitatively more important substrates than free amino acids [122].

3.   Genetic engineering and plant breeding for food purposes is a question of the future so far.

4.   The covalent incorporation of amino acids by peptide bond is very effective because the covalently bound amino acids are stable during food processing, storage, and cooking and have better nutritional efficiency in passing through the gut.

Enzymatic protein modifications, that is, enzymatic hydrolysis and resynthesis, have been described for improving the functional properties and the nutritive values of food proteins [37,46,123].

In practice, mainly soybean protein, gluten, and milk proteins have been nutritionally improved by incorporation of essential amino acids (Leu, Lys, Met) into the peptide chains. Amino acids to be incorporated are generally used in the form of L-amino acid esters [27,124,46]. Relative biological values NPR and NPU (netto protein ratio and netto protein utilization) of soy protein isolate, soy protein isolate supplemented with free L-Met, and soy protein isolate supplemented with Met-enriched EPM product were measured (Fig. 2) in rat feeding experiments [135]. These results showed that the biological value of the soybean isolate balanced by adding that of methionine-enriched EPM was basically better than that of the soy protein isolate, due to the covalently enriched methionine in the peptide chain of the EPM product.

Other results suggest that enzymatic peptide modification by Met incorporation results in increasing the nutritive value of buffalo's milk proteins [124], too. The role of dogfish chymotrypsin has been also investigated as an available catalyst in covalent amino acid enrichment [27]. Both dogfish chymotrypsin and bovine chymotrypsin were capable of incorporating methionine ester into soy protein. However, the use of bovine enzyme resulted in higher Met incorporation than that of the fish enzyme. That means that bovine chymotrypsin was more efficient in covalent methionine enrichment into soy protein [27].

In vitro experiments carried out in physiological conditions showed that *digestibility* (by gastrointestinal enzymes) of the methionine-enriched protein was similar to that of casein substrate [125]. It can be concluded that the structural alterations induced by protease-catalyzed reactions exert no negative effect on the nutritional value of the modified protein. The enzymatic availability of methionine from the methionine-enriched product and from casein is similar. Compared with the liberation rates of all the other amino acids, from both protein substances the methionine liberation rate is very high. These results are in agreement with the results of previous enzymic digestion studies with gastrointestinal proteolytic enzymes. These have demonstrated the high liberation rate of methionine from native proteins. The EPM product is just as easily digestible by physiological proteases and peptidases as casein. Casein is known as a well digestible protein under physiological conditions. It may be concluded that the enzymatic modification process does not reduce the enzymatic digestibility of the bonds within the protein molecule.

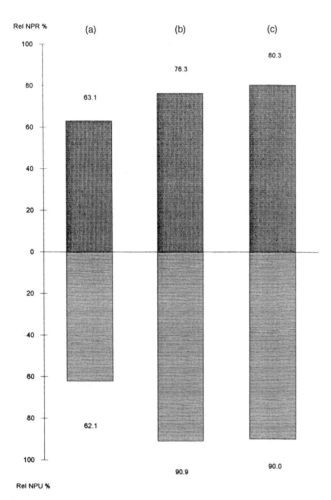

Figure 2 Relative biological values—NPR and NPU—of soy protein isolate and methionine supplemented soy protein isolates. (a) Soy protein isolate (Met + Cys →2.6g/16g N); (b) soy protein isolate supplemented with free L-methionine (Met + Cys →3.5g/16g N); (c) soy protein isolate supplemented with methionine-enriched EPM product (Met + Cys →3.5g/16g N).

D. Improvement of the Safety of Foods by SH-Containing Amino Acid Enrichment of Peptides

Recently there is a growing need to improve the quality and safety of food also via counteracting the formation of toxic compounds and/ or nutritional antagonists of the human organism. One of the best methods for this purpose is modifying the specific sites of food toxicants with site-specific reagents, such as SH-containing amino acids [126]. Because of their antitoxin and antioxidant effects, SH-containing amino acids are able to act as reducing agents, scavengers, and inducers of cellular detoxification [126,127,128]. With respect to biological utilization and safety it is important to note that only nutritious sulfur amino acids should be applied for food uses. However, sulfur amino acids, first of all methionine, are the limiting essential amino acids in most plant proteins. In order to improve the nutritional quality of foods, free L-methionine fortification is not the best method because of the toxic or antinutritional effect of high levels of free methionine in the diet [122,126].

Summarizing the above, proteolytic modification of proteins by covalent sulfur amino acid enrichment could be a suitable method for producing protein-based reducing agents, scavengers, and inducers of cellular detoxification, that is, for improvement in the safety of foods.

E. Modification of Functional Properties

Enzymatic modification of proteins has been elaborated also for desired modifications of functional properties of food proteins. The surface properties of a functional protein can radically change when the protein associates with other types of molecules possessing surfactant properties. The versatility of proteins is mainly due to their complex structure, and the variability of functional properties of proteins can be ascribed to differences in the structure of protein molecules, more exactly to varieties of their building amino acids [11]. The most important properties like surface activity and proteolytic degradability are basically influenced by compositional and structural features [129]. Enzymatic modification of food proteins may alter their charge frequency, their hydrophobicity, and their structure, thus modifying their functionality.

Protein-based ingredients contribute to the enhancement of food texture, flavor, and eye appeal [130]. Functionality of protein ingredients can be enhanced also by enzymatic modifications. The factors influencing the choice of an appropriate enzyme for improvement of functional properties of proteins are as follows: specificity of the enzyme, conformation of the protein, pH optimum, presence of activators and/or inhibitors, availability, thermostability, and financial causes.

The functional properties of proteins depend also on their structure and interactions with the environment. The functional properties of surfactants depend on their hydrophilic–hydrophobic balance, too. Protein chains modified by proteolysis, amino acid incorporation, and transpeptidation may display different functional properties. As milk proteins possess good surface activities [131], the question of the changes in the functional properties of the enzymatically modified protein products is of especial interest.

Enzymatic hydrolysis of proteins results in changing the molecular mass of the molecules. However, the effects of protein size on the hydrophobic behavior of amino acids are of great importance [132]. There is a meaningful relationship between hydrophobicity (which may affect the surface of the molecule) *and functionality* of food proteins [11,133] proposed using the term relative surface hydrophobicity. The proteolysis should be carefully limited for improving the functional properties of food proteins [147]. Mild hydrolysis improves functionality of proteins, while extensive hydrolysis depresses it [139,148].

Milk is rich in functional ingredients and also in surface-active peptides. These surface-active peptides are highly amphipathic. Amphipathic peptides influence functionality of milk proteins primarily by affecting their surface activity. Controlled hydrolysis of milk proteins results in markedly improved oil-holding capacity and solubility. Endogenous milk peptides affect the functionality of milk proteins [139].

Lee et al. [151] isolated a hydrophilic and a hydrophobic peptide each from the tryptic-chymosinic hydrolysate of α-casein. Interestingly, the emulsifying activities of these peptides depended on the pH range. Further work is needed in order to predict the relationship between structure and functionality of proteins. Casein and whey proteins were treated with trypsin, and functional properties of the hydrolysates were investigated [152]. The emulsifying activities of both tryptic hydrolysates were higher than those of the untreated protein.

With casein and whey proteins, hydrolysis depressed functionality with a number of enzymes. The result of this observation was that the activity of these enzymes could not be well controlled [139]. Controlled hydrolysis is available for producing a product with desired functionality.

Gallagher et al. [138] investigated the future application of two enzymes, bromelain and a bacillus protease (*Bacillus subtilis*) in the production of peptides from casein in point of view of the functional properties of the products. Bromelain action resulted in a hydrolysate with a great number of high-molecular-mass peptides; this may have improved the functional properties of a food product. The bacillus protease seemed to be more suitable for producing bitter peptides for future research and/or for future food.

The emulsifying activity of proteins [143] increased in the pH range of 2–9 [144] upon hydrolysis of soy proteins. Emulsifying capacity was also increased by hydrolysis of soy. Proteolysis of soy protein had no effect on foam stability [144,118], but hydrolysis of zein increased foam stability [145].

Gliadins, ovalbumin, bovine serum albumin, and α-lactoglobulin were treated by proteases at pH 10, and the authors [146] observed no deamidation due to enzymes, but the proteins were enzymatically digested. The enzymatically modified gliadins had very poor emulsifying and foaming properties, that is, their charge equilibrium was not changed.

Kimoto et al. [153] investigated the effect of proteinases from *Grifola frondosa* on gelling of egg white, and stated that no proteinase alone could prevent ovalbumin and egg white gelling, but a mixture of three proteinases prevented gelling. Proteinases deriving from *Grifola frondosa* would thus appear to degrade egg white protein and prevent its gelling.

Practically *all types of enzymatic modifications* of peptides *and* proteins are suitable processes for modification of viscoelastic [140] or other *functional properties* of food proteins. A special enzymatic technique (EPM) with covalent amino acid enrichment has also recently been elaborated for modification of the functionality of food proteins [134,135]. The basic changes in the functional properties of proteins are due very probably to two different factors:

1. Transpeptidation, which results in changes of covalent bonds within the peptide chains
2. Amino acid enrichment of the modified peptides, which leads to a favorable change in the structure of the proteins investigated.

The enzyme-catalyzed reaction should be carried out very carefully, without destroying existing functional properties important for the specific application of the protein [147].

Functional properties of some enzymatically modified and EPM-treated products of milk proteins [136] were determined as follows. An enzymatically prehydrolyzed commercial milk protein concentrate (SR) without further hydrolysis, and casein hydrolyzed by alcalase, α-chymotrypsin, and papain, respectively, were used as substrates in the EPM reaction. The concentration of the hydrolysates was 20% w/v in the EPM reactions. A methionine methyl ester hydrochloride/substrate ratio of 1:5 was used for incorporating this amino acid. After incubation, the products with methionine incorporation were simultaneously dialyzed for 2 days through a cellophane membrane against distilled water. The nondialyzable fractions and the EPM products without amino acid enrichment were freeze-dried. Covalent methionine incorporation in the EPM products with amino acid enrichment was verified by exopeptidase hydrolysis of the protein chains. The functional properties of the different EPM products are summarized in Table 1. An important functional property of proteins and/or peptide mixtures is their emulsifying behavior. This is highly influenced by the molecular structure, the position and ratio of hydrophobic-hydrophilic amino acids. Emulsion activity was found to be low (34.0) for casein, and the values determined for enzyme hydrolyzed and modified products were in general even lower. The papain hydrolysate, sample H3, showed here a different behavior as well; this was the one of the sample series that had the highest EAI value (43.0). The emulsion stability of the enzymatically modified products displayed tendencies quite opposite to the values of emul-

Table 1 Functional Properties of Casein and Enzymatically Modified Proteins[a,b]

Sample	EAI m²/g	ESI h	FC %	FS %/min.	Water-binding capacity	S_o
H1	30.0	0.6	450	5.6(6′)	0	270
EPM 1/a	19.6	1.6	213	0 (2′)	0.28	391
H2	12.0	2.5	175	0 (1,5′)	0	203
EPM 2	13.7	2.2	330	3.5(10′)	0	203
EPM 2/a	20.3	1.7	337	0.7(15′)	0.52	224
H3	43.0	0.4	450	0 (4′)	1.29	242
EPM 3	25.5	1.7	50	0 (0.5′)	0.80	146
EPM 3/a	21.7	1.8	45	9 (2′)	1.12	77
SR	39.0	0.2	413	3 (5′)	1.32	624
SR H4	31.0	1.4	113	0 (1.5′)	0	179
EPM 4	17.0	2.5	10	0 (1′)	0.31	117
EPM 4/a	18.8	1.4	119	2.2(2′)	0.96	179
Casein	34.0	0.2	421	5.3(10′)	1.84	475

[a] $S_o \to$ surface hydrophobicity; EAI $\to$ emulsifying activity; ESI $\to$ emulsifying stability; FC $\to$ foam capacity; FS %/(min.) $\to$ foam stability.
[b] H1, H2, H3: hydrolysates of casein by alcalase, α-chymotrypsin, and papain respectively; SR: enzymatically prehydrolysed commercial milk protein concentrate; SR H4: α-chymotryptic hydrolysate of SR; EPM 1/a: Met-enriched EPM product of H1 (enzyme catalyst: alcalase); EPM 2: EPM product of H2 without amino acid enrichment (enzyme catalyst: α-chymotrypsin); EPM 2/a: Met-enriched EPM product of H2 (enzyme catalyst: α-chymotrypsin); EPM 3: EPM product of H3 without amino acid enrichment (enzyme catalyst: papain); EPM 3/a: Met-enriched EPM product of H3 (enzyme catalyst: papain); EPM 4: EPM product of SR H4 without amino acid enrichment (enzyme catalyst: α-chymotrypsin); EPM 4/a: Met-enriched EPM product of SR H4 (enzyme catalyst: α-chymotrypsin).

sion activity. The ESI values were higher than the initial casein ESI values in all samples. In sample H3, the papain hydrolysate, with the highest EAI value, a relatively lower emulsion stability value (ESI = 0.4) was measured. Four samples of the casein-based products modified with α-chymotrypsin produced significant emulsion stability. These were the following: H2, EPM2, EPM2/a, and EPM4. Their ESI-values were 2.5, 2.2, 1.7, and 2.5 respectively.

Kato et al. [38] observed that with proteins of great molecular masses, there was a definite relation between surface hydrophobicity and emulsifying behavior. This observation could not be extended [136] to the proteolytic hydrolysates of casein and EPM products. Ludwig and Ludwig [137] also could not support the observations of Kato et al. [38] by their studies carried out with the enzymic modification of *Vicia faba* proteins. Nevertheless, with the peptide mixtures tested, another specific property merits attention: the ESI values of our enzyme-modified products tested are very high. ESI = 2.2 or ESI = 2.5 are such high indices that even proteins of great molecular masses do not always have emulsion stability. The foam capacity of the samples H1 and H3 was slightly higher than that of the initial protein. The foam capacity of the sample H1 was nearly similar to that of the casein. A significantly higher value of the foam stability was shown only by one of the casein samples, sample EPM 3/a. It is a very interesting result that the enzyme hydrolysates and enzymatically modified products of milk proteins do not produce high foam capacity to compare to that of the egg protein, but their low capacity is of a stable character. This essential increase in the water binding capacity of the Met-enriched samples is a direct consequence of the hydrophobic character of the methionine content built into the peptide chains. The hydrophobic character (S_o) determined by the soluble part of the samples indicates high values for the substrate proteins casein and SR, and for one of the products, the sample EPM 1/a, a Met-enriched product prepared by alcalase catalysis.

Arai et al. [141] described a particular enzymatic reaction for producing a surface-active protein. A highly hydrophobic amino acid was covalently bound to a hydrophilic protein in an enzyme-catalyzed process for this purpose. The covalent attachment of L-Leu *n*-alkyl ester to gelatin in the presence of papain as catalyst resulted in a proteinaceous surfactant [141,142] with very good emulsifying properties.

A surface-active macropeptide (MW 20,000) was produced as well by Toiguchi et al. [149] from succinylated α_{S_1}-casein by modification with papain. The reaction was carried out in the presence of L-leucine *n*-dodecyl ester. This was covalently bound to the peptide during the papain catalyzed reaction, and thus a macropeptide with amphiphilic structure was produced. Enzymatic modifications are very suitable reactions for producing proteinaceous surfactants via altering surface activity of the proteins.

Proteinaceous surfactant could be prepared by changing the ratio of the lipophilic–hydrophilic regions of a protein as well. Watanabe and coworkers [150] have succeeded in improving emulsifying and whipping properties by covalent attachment of a lipophilic leucine ester to gelatin in the presence of papain as catalyst. The authors suggested these papain-modified gelatin-based proteinaceous surfactants as ingredients in foods. This surfactant was applicable for the preparation of mayonnaise of a fine emulsion having favorable hardness and adhesiveness. This surfactant was usable also in bread making.

Because food products require proteins with special functional properties, water-soluble zein was produced using a dual-phase, sequential enzymatic modification and ultrafiltration. A significant improvement in certain functional properties of zein was observed upon this enzymatic modification in organic solvents [154].

Deamidation of proteins is important for improving functional properties of the product under mild reaction conditions. But enzymatic deamidation of proteins has not had real attention until recently. Kato et al. [155] developed a method for enzymatic deamidation of food proteins by treatment with proteases at pH 10. Salt and disulfide reducing agents have little effect on soy protein deamidation. Heat treatment and proteolysis of soy proteins are the major factors affecting deamidation [156].

The improvement of functional properties can be obtained by the help of enzyme modification with the directed change of the size and composition of peptides or of their terminal sequence. Several of these proteolytic modification reactions could be used even on a commercial scale in the food industry.

F. Production of Proteins with Antifreeze Character

The significance of preparing antifreeze emulsions is of great importance in point of view of biological systems and cold-sensitive foods.

Arai et al. [141,142] developed an enzymatically modified gelatin with great surfactancy. The hydrophobic L-leucine n-dodecyl ester was incorporated into the chains of the hydrophilic gelatin in the presence of papain as catalyst. The enzymatically modified gelatin (EMG-12) can act as an effective antifreeze protein. EMG-12 is able to regulate ice nucleation both in emulsion and in dispersion systems. This EMG-12 was used [157,158] in different concentrations to prepare antifreeze emulsions. This EMG-12–based antifreeze emulsion can maintain its unfrozen state even in the presence of silver iodide crystals added as ice nuclei. EMG-12 is highly surface active. Moreover, it is a bifunctional surfactant. It is able to produce an oil–water emulsion with very good functional properties and is useful as a very effective cryoprotectant. These types of surfactants and antifreeze dispersions [202] are very important in preserving living systems and for food and medical purposes.

G. Alteration of Biological Activities

1. Antigenic (Allergenic) Character of Proteins

Food allergy has been known for about two and a half thousand years. More than four hundred years before Christ, Hippocrates recorded the observation that "milk could cause special illnesses." Four hundred years later, Lucretius stated that "one man's food may be another man's poison."

In our day a great many researchers make efforts to solve the mystery of food allergy and to produce nonallergenic proteins for food and medical uses. Food allergy is an adverse reaction to a food or food component (mainly a protein) involving reactions of the body's immune system [159,160].

Patients suffering from food allergy have to avoid the allergenic components; they need hypoallergenic foods [161,162,163]. There are some nonenzymatic treatments producing hypoallergenic food proteins, such as heat treatment [164] and selective elimination of the major allergens from a protein mixture [165]. The only way of producing true hypoallergenic formulas is to modify the structure, the conformation, of the protein sufficiently, that is to reduce basically the quantity of the antigenic protein(s). The enzymatic modification of proteins is one of the best ways to reduce their antigenic character. During the cleaving of the polypeptide chain in enzymatic modification, the antigenic architecture of the protein molecule collapses. While sequential epitopes may survive the steps of enzymatic hydrolysis, epitopes are rapidly altered depending on the actual conformation of the protein. Protein hydrolysates for hypoallergenic formulas should have very special characteristics [166]: a great number of di- and tripeptides for quick absorption, a content of free amino acids below a given level, a limited quantity of bitter peptides, and high nutritional value. One of the most important things is to ensure that oral administration of the hypoallergenic formula does not induce sensitization to other proteins or residual peptides [167].

Proteins of several foods have been identified as common allergens: milk, legumes, eggs, cereals, seafood, etc. However, many patients suffering from a given food protein allergy are often sensitive also to other food proteins [188,189,190,191].

The identification of food allergens is mainly based on clinical observations before and after administration of an allergen-free diet to patients suffering from a given allergy.

Because of its absence in human milk, β-lactoglobulin is considered to be one of the major allergenic proteins of cow's milk [168]. Matsuda and Nakamura summarize the major food allergens along with their structural and immunological properties [169]. According to their collection, potent allergens in cow's milk are α_{S1}-casein (23 KDa), β-lactoglobulin (18 KDa), and Maillard adducts (amino-carbonyl reaction products between lactose and protein amino groups). The egg white's potent allergens are ovalbumin (43 KDa) and ovomucoid (28 KDa). Soybean has also several potential allergenic proteins such as glicinin (320–360 KDa), 2S-globulin (a mixture of low-molecular-weight proteins including trypsin inhibitors), and a 32 KDa allergen. There is a 16 KDa allergen in rice (belonging to the amylase/trypsin inhibitor family). It is also concluded that no common universal structural characteristic of the allergenic molecules has been found so far. However, it may be possible to eliminate the allergenic components or to reduce their level by several new food-processing techniques, first of all by enzymatic modification processes.

Proteolysates of proteins have been used for more than a half century for different nutritional purposes [170]. However, one of the most important goals is to produce protein hydrolysates or enzymatically modified proteins for infants or patients suffering from food allergy. The immunogenicity and allergenicity of protein hydrolysates have been evaluated by several authors [161,171,172]. The actual antigenicity of a protein can be followed by immunochemical methods. The functional properties of the enzymatically hydrolyzed and modified proteins have also been investigated [148,152,173,174].

Enzymatic modification, heat treatment (as an enzyme inactivating step) and separation processes result in a significant reduction of the allergenic character of the proteins. However, there are still problems in obtaining peptides of lower allergenicity. The first step is the selection of enzymes that attack the antigenic site of the protein. The heat treatment of food proteins leads to the loss of conformational epitopes during denaturation, but proteolysis may lead to the loss of sequential epitopes [178]. Controlled enzymatic hydrolysis can alter the surface character of proteins; it may destroy epitopes and thus may reduce the allergenicity of proteins or peptides. Therefore the length of peptide chains in a proteolytic hydrolysate is very important for producing such physiologically optimal hydrolysates as peptide-based hypoallergenic infant formulas [122].

Ultrafiltration treatment of enzymatic hydrolysates can further reduce the immunoreactivity of whey protein in vitro [179]. Nakamura and coworkers [176,177] stated that combinations of hydrolysis and membrane treatment (microfiltration or ultrafiltration) result in a desirable hypoallergenic peptide. The antigenicity of the fractions decreased with the decreasing of the pore size of the ultrafiltration membrane.

Proteolytic hydrolysates are increasingly used with patients suffering from protein hypersensitivity. Both the "first-generation" hydrolysates of casein and the "second generation" whey protein hydrolysates are highly hydrolyzed. Recently, "third generation," less degraded, whey protein hydrolysates have been described [122]. Infant formulas based on cow's milk protein have been widely used as supplements or substitutes [180]. However, bovine milk proteins might cause allergies in infants. Asselin et al. [181,182] suggested that allergenicity associated with α-lactalbumin and β-lactoglobulin could be reduced by α-chymotryptic hydrolysis. The most common problem in the enzymatic hydrolysis of milk proteins was the formation of peptides with bitter taste [184,185]. The study of the antigenic character and selected functional and physicochemical properties of extensively hydrolyzed bovine casein showed that [194] antigenicity loss occurred during extensive pancreatic hydrolysis.

The allergenic character of cow's milk casein, its EPM products with and without methionine enrichment, and the two fractions (selected by FPLC) of the Met-enriched EPM product were also investigated (Fig. 3). The allergenic character of the enzymatically modified products of cow's milk casein shows favorable changes: the allergen-

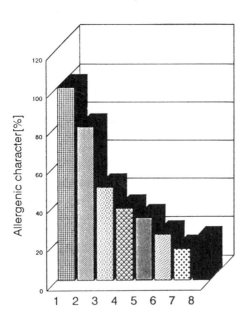

Figure 3 Allergenicity of cow's milk casein and of enzymatically modified cow's milk casein products. (1) Cow's milk casein; (2) α-chymotryptic hydrolysate of casein; (3,4) EPM products without Met enrichment; (5) EPM product with Met incorporation; (6,7) FPLC fractions of EPM products; (8) commercial hypoallergenic formula.

icity of the enzymatically hydrolyzed product (sample 2) has been significantly decreased in comparison to the untreated casein (sample 1), presumably because of the cleavage of a great number of peptide bonds. The EPM products and their selected fractions (samples 3–7) showed a more pronounced decrease in the allergenic character. These findings might well be due to transpeptidation processes in the course of EPM.

Pepsin and α-chymotrypsin were found to be the most efficient enzyme pair to reduce the allergenicity of whey proteins. Jost et al. [183] proposed a combination of selective proteolysis and heat treatment. Nakamura and coworkers [176,177] also investigated the antigenicity of whey protein hydrolysates prepared with 13 commercial proteases. The authors aimed at producing peptides with both good sensory properties (low bitterness) and hypoallergenicity. Hydrolysis with papain W-40 proved to be the most effective process for reducing the antigenicity of α-lactoglobulin. It is believed that using a combination of various proteases is a good method to develop hydrolysates for hypoallergenic formulas.

Pahud et al. [179] found that a tryptic hydrolysate of whey protein and a fraction of this hydrolysate obtained after ultrafiltration were devoid of the effect of sensitizing guinea pigs by the oral route. Nakamura et al. [180] investigated the antigenicity of whey protein hydrolysates prepared with a combination of two proteases, papain W-40^R and an endo-type protease of various bacterial origins. Their results showed that the antigenicity was significantly decreased with the combination of two kinds of enzymes. However, combined hydrolysis at high temperatures seemed to be a suitable method for producing peptides of low bitterness and low antigenicity from whey proteins.

The allergenic character (Fig. 4) of the α-chymotryptic and α-chymotryptic–tryptic hydrolysate of buffalo's milk proteins and their α-chymotryptic EPM products with different Met enrichments was significantly reduced compared to the unmodified buffalo's milk proteins [105]. The most essential decrease in the allergenic activity was found in the α-chymotryptic EPM samples with Met enrichment, probably due to the alteration of the sequential and conformational determinants of the proteins according to the transpeptidation and co-

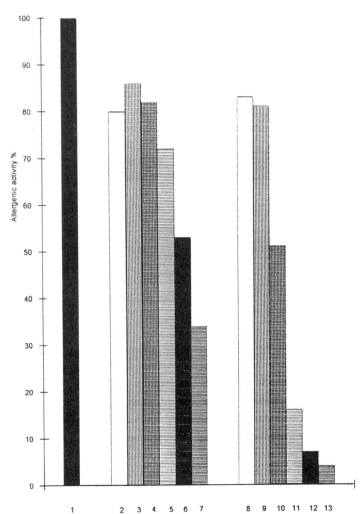

Figure 4 Allergenic activity of the EPM products produced from different proteolytic hydrolysates of buffalo's milk proteins. (1) Buffalo's milk proteins (control); (2) EPM product produced from the chymotryptic hydrolysate without amino acid enrichment; (3,4,5,6,7) (α-chymotryptic) EPM products with different Met enrichments; (8) α-chymotryptic product produced from the α-chymotryptic and tryptic hydrolysate (without amino acid enrichment); (9,10,11,12,13) α-chymotryptic EPM products with different Met enrichments produced from a peptic and tryptic hydrolysate of buffalo milk proteins. The allergenic activity of the samples was measured *in vitro* by competitive indirect ELISA.

162

valent amino acid incorporation. The allergenic character of the peptic and peptic-tryptic hydrolysate of buffalo's milk proteins and their peptic EPM products (Fig. 5) was also investigated [105]. The allergenicity of all samples was reduced in comparison with the untreated samples. The degree of this reduction, however was not as high as in the case of the α-chymotrypsin–catalyzed products (Fig. 4).

For the enzymatic decomposition of rice proteins a suspension of Actinase AS (at 37°C for 24 h) was used. As a result of the process, the major allergenic globulin of rice was decomposed. In Japan there are many examples of rice-associated allergic diseases, because rice is consumed as a staple foodstuff [187]. Matsuda and Nakamura [169] and Izumi et al [192] characterized the major allergens in rice. Watanabe et al. [193] proposed a method for producing hypoallergenic rice for patients suffering from rice allergy.

Wheat-associated allergy is very serious, but allergenic proteins of wheat and/or their sequential and conformational epitopes have not been specified. Watanabe and coworkers [186] have tried to develop controlled enzymatic treatment for producing hypoallergenic wheat flour. Food allergens are sometimes characterized by stability against digestive enzymes. Therefore enzymes used in this work were different from gut digestive enzymes; they were molsin, actinase (neutral proteinase), alkaline proteinase, and furthermore collagenase, and transglutaminase. Treatments with actinase, collagenase and transglutaminase decreased the allergenic character of the protein fractions. The actinase-treated product contained rather low-molecular-weight proteinaceous components, but the collagenase- or transglutaminase-treated product retained some high-molecular-weight proteinaceous components. These results suggest that, for food processing, the latter product may be preferable to the product resulting from treatment with actinase.

Soy proteins are strong antigens. The gut wall is the site of food allergy generation. Usually the loss of sufficient barriers permits antigens to cross the absorbing surface and to contact underlying immune structures, while modifying the bioavailability of proteins. The goal of recent works [123,190,195] was a comparative study on the bioavailability of different soy products: raw soybean, heat-treated soy protein concentrate, soy protein isolate, methionine-supplemented soy protein isolate, proteolytic hydrolysate of soy protein isolate, and enzymati-

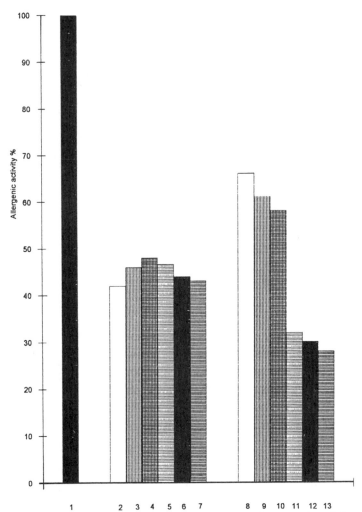

Figure 5   Allergenic activity of the EPM products produced from different proteolytic hydrolysates of buffalo's milk proteins. (1) Buffalo's milk proteins (control); (2) peptic EPM product produced from a peptic hydrolysate (without amino acid enrichment); (3,4,5,6,7) peptic EPM products with different EPM enrichment from the peptic hydrolysates; (8) peptic EPM product produced from the peptic and tryptic hydrolysate (without amino acid enrichment); (9,10,11,12,13) peptic EPM products with different Met enrichment produced from the peptic and tryptic hydrolysate of the buffalo milk proteins. The allergenic activity of the samples was measured *in vitro* by competitive indirect ELISA.

cally modified (EPM) products of soybean isolate with and without amino acid enrichment. Allergenic characters were not significantly reduced in physiocochemically treated soy proteins. However, the potential allergenic activity significantly dropped in hydrolysate of soy isolate owing to the cleavage of a great number of peptide bonds. The most significant decrease in allergenic activities was measured in EPM products owing to the modification of the structure of sequential and conformational determinants by transpeptidation and covalent attachment of methionine to the protein chains (Fig. 6).

Antigenicity of two enzymatic hydrolysates of whey protein and casein were studied in rats and guinea pigs by Boza et al. [166]. Both hydrolysates contained free amino acids in small quantities for preventing an increase in osmolarity of the formula. The potential antigenicity of the whey protein hydrolysate was reduced 10^3 times, whereas that of the casein hydrolysate was reduced 10^4 times compared to that of the casein as measured in vitro by ELISA. The in vivo allergenicity tests of this hydrolysate of whey protein obtained without ultrafiltration showed some positive reactions in the systemic and passive cutaneous anaphylaxis tests. The possible interpretation of this fact is that antigens capable of provoking an allergic response were still present in the hydrolysate. This type of formula should not be used for infants with food allergies. However, this casein hydrolysate showed hypoantigenicity and did not induce oral sensitization. The authors stated that further studies were needed to elucidate the clinical efficiency of this casein hydrolysate in patients suffering from cow's milk allergy.

Allergenicity of proteins of milk and various (heat-treated, fermented, and enzymatically modified) milk products was determined in vitro by ELISA [203]. Food processing like heat treatment or fermentation does not reduce the allergenic character of the proteins. The enzymatic modification, however, showed a significant reduction in the allergenicity of milk proteins (Fig. 7).

That the allergenic character of proteins could be decreased by this enzymatic method is due probably to the following factors:

1. Proteolysis reduces the immunogenic character of the modified proteins.
2. The modification of the structures of sequential and conforma-

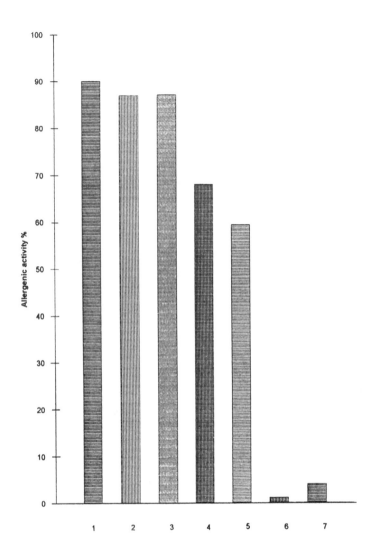

Figure 6 Potential allergenic activity of different soy protein products. Potential allergenic activity of competitive soy proteins as percentage of maximal binding activity at the concentrations of competitive antigens of 10^2 μg/mL. (1) Raw soybean, dehulled; (2) heat-treated extrusion soy protein concentrate; (3) soy protein isolate; (4) proteolytic hydrolysate of soy protein isolate; (5) EPM (1) product of soybean isolate without amino acid enrichment; (6) EPM (2) product of soybean isolate with methionine enrichment; (7) EPM (3) product of soybean isolate.

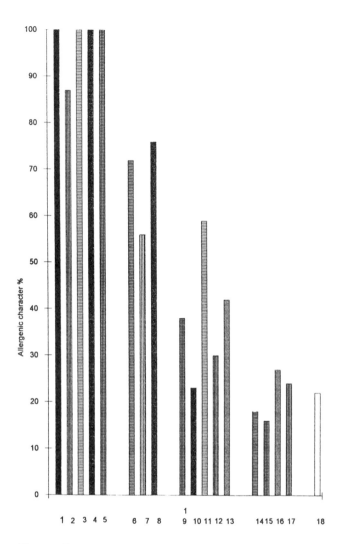

Figure 7   Allergenic character of products obtained from cow's milk proteins by food processing or enzymatic modifications. (1) Cow's milk; (2) Na-caseinate; (3) kefir; (4) yogurt; (5) cheese; (6,7,8) α-chymotryptic, tryptic, and peptic hydrolysates of casein, respectively; (9,10,11) α-chymotryptic tryptic, and peptic EPM products of casein, respectively; (12,13) α-chymotryptic and tryptic EPM products of casein, respectively, with methionine enrichment; (14,15) fractions of α-chymotryptic EPM products of casein; (16,17) fractions of peptic EPM products of casein; (18) commercial hypoallergenic formula.

tional determinants by transpeptidation leads to a favorable change in the antigenicity of protein chains.

3. Covalent incorporation of designed amino acids modifies the structure of the determinant groups of the proteins and prevents the formation of new determinants.

Thus the allergenicity of proteins may be satisfactorily reduced by a designed enzymatic modification of immunogenicity and antigenicity.

2. Antinutritive Properties of Proteins

Legume seeds contain a number of antinutritive components, mainly lectin and trypsin inhibitors. The efficiency of the nutritional utilization of diets containing soybean is well below that expected on the basis of chemical composition [196]. In order to reduce the extent of this constraint, at present, all soy products go through expensive heat treatment or other processing procedures that can lead to losses of essential amino acids and to the production of toxic by-products. It is hoped that an effective strategy will help improve the nutritional value of soy proteins [197,198].

Recently an adequate analytical method was developed [199] for selecting enzymes capable of reducing or eliminating the antinutritive activity of proteinaceous compounds. The effects of three enzyme preparations were examined on two soy products with respect to their capability to eliminate different antinutritional factors and to reduce antigenicity of the soy product.

The soy storage proteins (glycinin β-conglycinin) and an antigenic soy protein (of low molecular mass) were degraded by the proteolytic process and an apparent loss of antigenicity was observed. However, the proteolytic effect on soy lectin and Kunitz trypsin inhibitor was obviously lost, although some minor decrease in immunoreactivity could be observed.

With the enzymatic modification of soy albumin (EPM soy albumin), its nutritional utilization was improved. Animals fed soy albumin lost weight during the 10-day experimental period. While the rats were fed EPM-soy albumin, they kept their body weight constant or improved it slightly [198]. During the process of the enzymatic modification of soy albumin, the antinutrients (Kunitz trypsin inhibitor, Bowman–Birk trypsin inhibitor, and soybean agglutinin) of soybean were partially eliminated. The protein band representing the lectin content of soy albumin disappeared from the SDS-PAGE pattern of the EPM-soy albumin.

The immunological activity of the soybean agglutinin content of soy albumin samples was also investigated. After an SDS-PAGE separation of the soy albumin samples, the proteins were electrophoretically transferred to a membrane and immunologically detected. Both the untreated soy albumin and the different proteolytic hydrolysates of soy albumin showed considerable soybean agglutinin activity. However, the sample of EPM-soy albumin did not indicate any significant activity against antisoybean agglutinin rabbit IgG antibody.

The investigation of the biological/functional activities of soy trypsin inhibitors showed unexpected results (Fig. 8). While BBI activity did not change during the enzymatic treatments, KTI activity was significantly reduced in the peptic hydrolysate and EPM product of soy albumin.

3. Antimutagenic Effect of Enzymatically Modified Proteins

Mutagenesis is known to be the very first crucial step of multistage carcinogenesis, in which DNA damaged by genotoxic agents finally leads to the transformation of a normal cell into a cancerous one.

The antimutagenic effect of several enzymatically modified (EPM) compounds was studied by the plate incorporation method of the Ames Salmonella Mutagenicity assay. Using 9-aminoacridine as a positive control for the TA97a strain and sodium azide for the TA100 strain, the samples were tested for their mutagenic, preventive, and curing effects. Even though these EPM compounds showed no significant preventive or curing effect against mutagens, they reduced spontaneous mutation rate in the TA97a strain, indicating that these EPM products are possible antimutagenic agents [200].

Three samples showed a special effect in reducing the number of spontaneous mutations. These samples were (1) an α-chymotryptic EPM product with methionine enrichment produced from an α-chymotryptic hydrolysate of casein; (2) a histidine-enriched EPM product

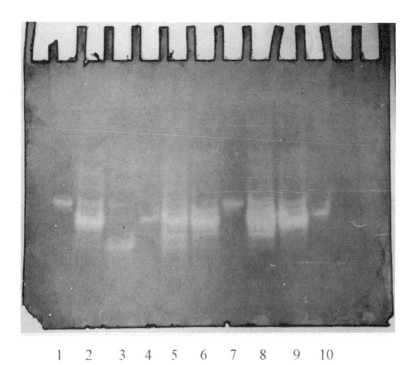

1   2   3   4   5   6   7   8   9   10

Figure 8   Reaction of trypsin inhibitors and enzymatically modified soy albumin samples with trypsin after native PAGE. Samples run were (1) Kunitz trypsin inhibitor (KTI); (2) α-chymotryptic hydrolysate of soy albumin; (3) α-chymotryptic EPM of soy albumin; (4) Bowman–Birk inhibitor (BBI); (5) peptic hydrolysate of soy albumin; (6) peptic EPM of soy albumin; (7) KTI; (8) proteolytic hydrolysate of soy albumin; (9) EPM product of soy albumin; (10) BBI. The bands were visualized by exposure to trypsin and its substrate *N*-acetyl-DL-phenylalanine β-naphthyl ester followed by negative staining with tetrazotized *o'*-dianisidine-$ZnCl_2$ complex.

obtained in the presence of α-chymotrypsin from a tryptic–α-chymotryptic hydrolysate of casein; (3) an EPM product obtained with covalent Met enrichment in the presence of α-chymotrypsin as catalyst from an α-chymotryptic–tryptic hydrolysate of casein.

VI. FUTURE OF THE ENZYMATIC MODIFICATIONS OF FOOD PROTEINS

What is being accomplished in food science and nutrition in relation to modified proteins for our future food?

A new concept and expectation have been developed over the past several years with respect to the protein safety and quality of food. We need proteins with specific physiological functions [201] and with a particular protein quality [1]. The most important of these desired particular properties of food proteins are safeguarding health, boosting the immune system, antagonizing cancer, aiding brain function, reducing stress, improving the health of the heart, etc.

Proteins and modified proteins will play a major role in producing foods in the future.

REFERENCES

1. E. C. Henley and J. M. Kuster, *Food Techn.* 4: 74 (1994).
2. M. Kajtár, in *Változatok négy elemre* (Variations on four elements), Gondolat, Budapest, 1984.
3. A. Pour-El, in *Functionality and Protein Structure* (R. F. Gould, ed.), American Chemical Society, Washington, D.C., 1979, p. ix.
4. R. E. Feeney and J. R. Whitaker, in *Protein Tailoring for Food and Medical Uses* (R. E. Feeney and J. R. Whitaker, eds.), Marcel Dekker, New York and Basel, 1986.
5. H. E. Swaisgood, S. X. Chen, S. Oh, and G. L. Catignani, in *Protein Structure-Function Relationships in Foods* (R. Y. Yada, R. L. Jakcman, and J. L. Smith, eds.), Blackie Academic and Professional, London, 1994.
6. J. R. Whitaker, in *Food Proteins* (R. E. Feeney and J. R. Whitaker, eds.), American Chemical Society, Washington, D.C., 1977, pp. 95–155.
7. J. S. Hamada, in *Biochemistry of Food Proteins* (B. J. F. Huson, ed.), Elsevier Applied Science, London and New York, 1992, p. 249.
8. J. E. Kinsella, S. Damodaran, and B. German, in *New Protein Foods*, Vol. 5. (A. M. Altschul and H. I. Wilcke, eds.), Academic Press, New York, 1985, pp. 107–178.
9. F. F. Shih, *J. Food Sci.* 55(1): 127 (1990).

10. M. Watanabe and S. Arai, in *Biochemistry of Food Proteins* (B. J. F. Hudson, ed.), Elsevier Applied Science, London and New York, 1991, pp 271–305.
11. S. Nakai, *J. Agric. Food Chem. 31*: 676 (1983).
12. I Nir, A. Feldman, and N. Garti, *J. Food Sci. 59*(3): 606 (1994).
13. S. Nakai and E. Li-Chan, *Crit. Rev. Food Sci. Nutr. 33*(6): 477 (1993).
14. L. Lorand and S. Conrad, *Mol. Cell. Biochem. 58*: 9 (1984).
15. K. Ikura, T. Kometani, R. Sasaki, and H. Chiba, *Agric. Biol. Chem. 44*(12): 2979 (1980).
16. M. Motoki and N. Nio, *J. Food Science 48*: 561 (1983).
17. M. Motoki, N. Nio, and K. Takinami, *Agric. Biol. Chem. 51*(1): 237 (1987).
18. K. Ikura, M. Yoshikawa, R. Sasaki, and H. Chiba, *Agric. Biol. Chem. 45*: 11): 2587 (1981).
19. J. S. Hamada, *Food Sci. and Nutrition 34*(3): 283 (1994).
20. J. S. Hamada, *J. Agric. Food Chem. 40*(5): 719 (1992).
21. J. S. Hamada and W. E. Marshall, *J. Food Sci. 53*: 1132 (1988).
22. B. P. Gill. A. J. O'Shaughnessey, P. Henderson, and D. R. Headon, *Irish J. Food Sci. Technol. 9*: 33 (1985).
23. N. F. Campbell, F. F. Shih, and W. E. Marshall, *J. Agric. Food Chem. 40:* 403 (1992).
24. K. Seguro and M. Motoki, *Agric. Biol. Chem. 53*(12): 3263 (1989).
25. J. R. Whitaker, in *Protein Tailoring for Food and Medical Uses* (R. E. Feeney and J. R. Whitaker, eds.), Marcel Dekker, New York, 1986, pp. 41–47.
26. C. Pallavicini, J. W. Finley, W. L. Stanley, and G. G. Watters, *J. Sci. Food Agric. 31*: 273 (1980).
27. M. Ramakrishna, H. O. Hultin, and M. T. Atallah, *J. Food Sci. 52*(5) (1987).
28. N. F. Haard, L. A. W. Feltham, N. Helbig, and E. J. Squires, in *Modification of Proteins. Food,* Nutritional and Pharmacological Aspects (R. E. Feeney and J. R. Whitaker, eds.), Advances in Chemistry Series 198, American Chemical Society, Washington, D.C., 1982, p. 223.
29. W. W. Sawjalow, *Z. Physiol. Chem. 54*: 119 (1907).
30. H. Borsook, *Adv. Protein Chem. 8*: 127 (1953).
31. L. M. Greenbaum and J. S. Fruton, *J. Biol. Chem. 226*: 173 (1957).
32. H. Determann, J. Heuer, and D. Jaworek, *Ann. Chem. 690*: 189 (1965).

33. S. Moore and W. H. Stein, *J. Biol. Chem. 211*: 893 (1954).
34. C. Tanford, *J. Am. Chem. Soc. 84*: 4240 (1962).
35. P. F. Fox, P. A. Morrissey, and D. M. Mulvihill, in *Developments in Food Proteins* (B. J. F. Hudson, ed.), Applied Science, London, 1982, pp. 1–60.
36. A. Halász and R. Lásztity, in *Use of Yeast Biomass in Food Production* (A. Arbor, ed.), Boston, CRC Press, 1991.
37. P. Lozano, D. Combes, and J. L. Iborra, *J. Food Sci. 59*(4): 876 (1994).
38. A. Kato, K. Komatsu, K. Fujimoto, and K. Kobayashi, *J. Agric. Food Chem. 33*: 931 (1985).
39. T. Nakano, M. Simatani, Y. Murakami, N. Sato, and T. Idota, *J. Jp. Soc. Nutr. Food Sci. 47*(3): 195 (1994).
40. S. K. Kim and E. H. Lee, *Han'guk Susan Hakhoechi 20*: 282 (1987).
41. J. S. Fruton, *Adv. Enzymol. 53*: 239 (1982).
42. H. Tauber, *J. Am. Chem. Soc. 73*: 1288 (1951).
43. T. Wieland, H. Determann, and E. Albrecht, *Liebigs Ann. Chem. 633*: 185 (1960).
44. H. Determann and R. Köhler, *Justus Liebigs Ann. Chem. 690*: 197 (1965).
45. M. Fujimaki, M. Yamashita, S. Arai, and H. Kato, *Agric. Biol. Chem. 34*: 1325 (1970).
46. M. Fujimaki, S. Arai, and M. Yamashita, *Adv. Chem. Ser. 160*: 156 (1977).
47. S. Arai, M. Yamashita, H. Kato, and M. Fujimaki, *Agric. Biol. Chem. 34*: 729. (1970).
48. M. Yamashita, S. Arai, M. Gonda, H. Kato, and M. Fujimaki, *Agric. Biol. Chem. 34*: 1333 (1970).
49. M. Yamashita, S. Arai, S. Tsai, and M. Fujimaki, *J. Agric. Food Chem. 19*: 1151 (1971).
50. B. V. Hofsten and G. Lalasidis, *J. Agric. Food Chem. 24*: 460 (1976).
51. V. Sciancalepore and V. Longone, *J. Dairy Res. 55*: 547 (1988).
52. S. J. Tsai, M. Yamashita, S. Arai, and M. Fujimaki, *Agric. Biol. Chem. 38*: 641 (1974).
53. S. J. Tsai, M. Yamashita, S. Arai, and M. Fujimaki, *Agric. Biol. Chem. 36*: 1045 (1972).
54. M. Yamashita, S. Arai, S. Tanimoto, and M. Fujimaki, *Agric. Biol. Chem. 37*(4): 953 (1973).
55. H. Determann, K. Bonhard, R. Köhler, and T. Wieland, *Helv. Chim. Acta 46*: 2498 (1963).

56. S. R. Noar and W. F. Shipe, *J. Food Sci. 49*: 1316 (1984).
57. J. H. Edwards and W. F. Shipe, *J. Food Sci. 43*: 1215 (1978).
58. G. Sukan and A. T. Andrews, *J. Dairy Res. 49*: 279 (1982).
59. A. T. Andrews and E. Alichanidis, *Food Chem. 35*: 243 (1990).
60. J. Horowitz and F. Haurowitz, *Biochim. Biophys. Acta 33*: 231 (1959).
61. M. Yu Gololobov, T. V. Antonova, and V. M. Belikov. *Die Nahrung 30*: 289 (1986).
62. Gy. Hajós and A. Halász, *Acta Alimentaria 11*(2): 189 (1982).
63. H. Delincée and Gy. Hajós, *Acta Alimentaria 13*: 309 (1984).
64. Gy. Hajós, F. Mietsch, and A. Halász, *Acta Alimentaria 17*(3): 265 (1988).
65. P. Lozano and D. Combes, *Appl. Biochem. Biotechn. 33*: 51 (1992).
66. Gy. Hajós, *Die Nahrung 30*: 418 (1986).
67. Gy. Hajós, in *Élelmiszerfehérjék módosítása proteázokkal* (Enzymatic peptide modification of proteins), Mezőgazdasági Kiadó, Budapest, 1986.
68. Gy. Hajós, A. Halász, and F. Békés, *Acta Alimentaria 18*(3): 325 (1989).
69. P. Lozano and D. Combes, *Biotechnol. Appl. Biochem. 14*: 212 (1991).
70. R. C. Hagan and R. Villota, *Food Chemistry 23*: 277 (1987).
71. F. J. Kézdy, S. P. Jindal, and M. L. Bender, *J. Biol. Chem. 247*: 5746 (1972).
72. M. Yamashita, S. Arai, Y. Imaizumi, Y. Amano, and M. Fujimaki, *J. Agric. Food Chem. 27*: 52 (1979).
73. M. Yamashita, S. Arai, Y. Amano, and M. Fujimaki, *Agric. Biol. Chem. 43*: 1065 (1979).
74. J. L. Iborra, P. Lozano, J. M. Obón, A. Manjón, and D. Combes, in *Profiles on Biotechnology* (T. G. Villa and J. Abalde, eds.), Servicio de Publicaciones Universitarias, Universidad de Santiago, Spain, 1992, pp. 477–487.
75. P. Lozano and D. Combes, *J. Sci. Food Agric. 62*: 245 (1993).
76. T. A. Andrews, Food Research Institute Reading; Report, April 1985.
77. K. Aso, M. Yamashita, S. Arai, and M. Fujimaki, *Agric. Biol. Chem. 38*: 679 (1974).
78. K. Aso, M. Yamashita, S. Arai, and M. Fujimaki, *J. Biochem. 76*: 341 (1974).
79. S. Eriksen and I. S. Fagerson, *J. Food Sci. 41*: 490 (1976).
80. H. Nötzold, H. Schafer, and E. Ludwig, *Die Nahrung 27*: 71 (1983).

81. H. Winkler, H. Noetzold, and E. Ludwig, *Die Nahrung 32*: 135 (1988).
82. Gy. Hajós, I. Éliás, and A. Halász, *J. Food Sci. 53*(3): 739 (1988).
83. Gy. Hajós, H. Nötzold, A. Halász, and E. Ludwig, *Acta Alimentaria 19*(1): 73 (1990).
84. D. Combes and P. Lozano, *Ann. N.Y. Acad. Sci. 672*: 409 (1992).
85. J. Léonil, D. Mollé, S. Bouhallab, and G. Henry, *Enzyme Microb. Technol. 16*: 591 (1994).
86. A. I. Virtanen, T. Laaksonen, and M. Kantola, *Acta Chem. Scand. 5*: 316 (1951).
87. B. Hahn-Hagerdal, *Enzyme Microb. Technol. 8*: 322 (1986).
88. T. Oka and K. J. Morihara, *Biochemistry* (Tokyo) *84*: 1277 (1978).
89. H. F. Gaertner and A. J. Puigserver, *Proteins Struct. Funct. Genet. 3*: 130 (1988).
90. H. F. Gaertner, N. Lupi, and A. Puigserver, *Le Lait 62*: 578 (1992).
91. H. Kise, K. Fujimoto, and H. Noritomi, *J. Biotechnol. 8*: 279 (1988).
92. M. Yu. Gololobov, V. M. Stepanov, T. L. Voyushina, I. P. Morozova, and P. Adlercreutz, *Enzyme Microbial Tech. 16*(6): 522 (1994).
93. H. Kise, A. Hayakawa, and H. Noritomi, *J.Biotechnol. 14*: 239 (1990).
94. A. F. Biagnini, H. Gaertner, and A. Puigserver, *J. Agric. Food. Chem. 41*: 1152 (1993).
95. K. Aso, S. Tanimoto, M. Yamashita, S. Arai, and M. Fujimaki, *Agric. Biol. Chem. 43*: 1147 (1979).
96. M. Yamashita, S. Arai, and M. Fujimaki, *J. Agric. Food Chem. 24*: 1100 (1976).
97. P. C. Lorenzen and E. Schlimme, *Milchwissenschaft 47*(8): 499 (1992).
98. M. Yamashita, S. Arai, K. Aso, and M. Fujimaki, *Agric. Biol. Chem. 36*: 1353 (1972).
99. D. V. M. Ashley, R. Temler, D. Barclay, C. A. Dormond, and R. Jost, *J. Nutr. 113*: 21 (1983).
100. J. C. Monti and R. Jost, *J. Agric. Food Chem. 27*: 1281 (1979).
101. M. R. Alecio, M. L. Dann, and G. Lowe, *Biochem. J. 141*: 49 (1974).
102. H. Aso, H. Kimura, M. Watanabe, and S. Arai, *Agric. Biol. Chem. 49*(6): 1649 (1985).
103. G. Anderson and P. G. Luisi, *Helv. Chim. Acta 49*: 488 (1979).
104. Gy. Hajós, T. Szarvas, and L. Vámos-Vigyázó, *J. Food Biochem. 14*: 381 (1990).

105. S. Hussein, É. Gelencsér, and Gy. Hajós, unpublished, 1995.
106. Zs. Demeczky, S. Hussein, and Gy. Hajós, *Acta Alimentaria* 23(1): 85 (1994).
107. A. Ferjancic-Biagini, H. Gaertner, and A. Puigserver, *J. Agric. Food Chem. 41*: 1152 (1993).
108. M. Yamashita, S. Arai, S. Kokubo, K. Aso, and M. Fujimaki, *J. Agric. Food Chem. 23*: 27 (1975).
109. S. Arai, M. Yamashita, and M. Fujimaki, *J. Nutr. Sci. Vitaminol. 22*: 447 (1976).
110. M. Yamashita, S. Arai, J. Matsuyama, H. Kato, and M. Fujimaki, *Agric. Biol. Chem. 34*: 1492 (1970).
111. J. F. Gordon, in *Proteins as Human Food*, (R. A. Lawrie, ed.), AVI, Westport, 1970, p. 328.
112. T. Matoba and T. Hata, *Agr. Biol. Chem. 36*: 1423 (1972).
113. B. Tchorbanov and I. Iliev, *Enzyme Microb. Technol. 15*: 974 (1993).
114. K. Surówka and M. Fik, *J. Sci. Food Agric. 65*: 289 (1994).
115. C. C. Bigelow, *J. Theor. Biol. 16*: 187 (1967).
116. M. Yamashita, S. Arai, S. Tanimoto, and M. Fujimaki, *Biochim. Biophys. Acta 358*: 105 (1974).
117. J. Adler-Nissen, *J. Agric. Food Chem. 24*(6): 1090 (1976).
118. G. Puski, *Cereal Chem. 52*: 655 (1975).
119. D. B. A. Silk, *Gut 15*: 495 (1974).
120. D. M. Matthews and S. A. Adibi, *Gastroenterology 71*: 151 (1976).
121. C. Kies and H. M. Fox, Abstracts of the 179th National Meeting of the American Chemical Society, Houston, TX, 1980.
122. A. D. Siemensma, W. J. Weijer, and H. J. Bak, *Trends Food Sci. Tech. 4*: 16 (1993).
123. Gy. Hajós, S. Hussein, and É. Gelencsér, in *Food Proteins: Structures and Functionality* (K. D. Schwenke and R. Mothes, eds.), VCH, Weinheim, 1993, pp. 82–86.
124. S. Hussein and Gy. Hajós, *Acta Alimentaria 22*(4): 351 (1993).
125. J. Noack and Gy. Hajós, *Acta Alimentaria 13*(3): 205 (1984).
126. M. Friedman, *J. Agric. Food Chem. 42*: 3 (1994).
127. G. Csomós and J. Fehér, in *Free Radicals and the Liver*, Springer-Verlag, Berlin, 1992.
128. Gy. Hajós, É. Gelencsér, A. Blázovics, and J. Fehér, unpublished, 1994.
129. E. H. Reimerdes. *Food Biotechnology 4*(1): 59 (1990).
130. A. Kilara. *Process Biochem.* 149 (October 1985).

131. J. E. Kinsella and D. M. Whitehead, in *Proteins at the Interfaces* (J. L. Busch and T. A. Hobett, eds.), American Chemical Society, Washington, D.C., 1987.

132. H. Meirovitch, S. Rackovsky, and H. A. Scheraga, *Macromolecules* *13*: 1398 (1980).

133. W. Melander and C. Horvath, *Arch. Biochem. Biophys. 183*: 200 (1977).

134. Gy. Hajós, É. Gelencsér, J. Bodnár, Cs. Sisak, and B. Mátrai, *Kém. Közl. 77*: 121 (1993).

135. Gy. Hajós, É. Gelencsér, S. Hussein, and M. Polgár, in *Food Industry: A Challenge to Biochemistry* (E. Czoboly, ed.), Hungarian Technical Cooperation Center Foundation, 1994, pp. 71–74.

136. I. Ludwig, W. Krause, and Gy. Hajós, *Acta Alimentaria*, submitted, 1995.

137. I. Ludwig and E. Ludwig, *Die Nahrung 29*(10): 949 (1985).

138. J. Gallagher, A. D. Kanekanian, and E. P. Evans, *Int. J. Food Sci. Tech. 29*: 279 (1994).

139. Z. U. Haque, *J. Dairy Sci. 76*: 311 (1993).

140. J. M. Chobert, C. Bertrand-Harb, and M. G. Nicolas, *J. Agric. Food Chem. 36*: 883 (1988).

141. S. Arai, M. Watanabe, and N. Fujii, *Agric. Biol. Chem. 48*(7): 1861 (1984).

142. S. Arai, M. Watanabe, and R. F. Tsuji, *Agric. Biol. Chem. 48*(8): 2173 (1984).

143. K. N. Pearce and J. E. Kinsella, *J. Agric. Food Chem. 26*: 716 (1978).

144. J. S. Hamada and W. E. Marshall, *J. Food Sci. 54*(3): 598 (1989).

145. A. A. Kossiakoff, *Science 240*: 191 (1988).

146. S. Bollecker, G. Viroben, Y. Popineau, and J. Gueguen, *Sciences des Aliments 10*: 343 (1990).

147. J. Adler-Nissen, in *Enzymatic Hydrolysis of Food Proteins*, Elsevier Applied Science, New York, 1986, pp. 9–29.

148. G. B. Quaglia and E. Orban, *J. Food Sci. 55*: 1571, 1619 (1990).

149. S. Toiguchi, S. Maeda, M. Watanabe, and S. Arai, *Agric. Biol. Chem. 46*(12): 2945 (1982).

150. M. Watanabe, A. Shimada, E. Yazawa, T. Kato, and S. Arai, *J. Food Sci. 46*: 1738 (1981).

151. S. W. Lee, M. Shimizu, S. Kaminogawa, and K. Yamauchi, *Agric. Biol. Chem. 51*: 161 (1987).

152. J. M. Chobert, C. Bertrand-Harb, M. Dalgalarrondo, and M. G. Nicolas, *J. Food Biochem. 13*: 335 (1989).

153. K. Kimoto, A. Hayashi, M. Kusama, T. Sugawara, and Y. Aoyagi, *J. Jp. Soc. Nutr. and Food Sci.* 47(1): 43 (1994).
154. A. Mannheim and M. Cheryan, *Cereal Chem.* 70(2): 115 (1993).
155. A. Kato, A. Tanaka, N. Matsudomi, and K. Kobayashi, *J. Agric. Food Chem.* 35: 224 (1987).
156. J. S. Hamada, F. F. Shih, A. W. Frank, and W. E. Marshall, *J. Food Sci.* 53: 671 (1988).
157. M. Watanabe, *Nippon Nogeikagaku Kaishi 61*: 482 (1987).
158. M. Watanabe, R. F. Tsuji, N. Hirao, and S. Arai, *Agric. Biol. Chem.* 49(11): 3291 (1985).
159. J. Brostoff and S. J. Challacombe, in *Food Allergy and Intolerance* (Bailliére Tindell, ed.), London, 1989.
160. C. D. May and S. A. Bock, *Allergy 33*: 166 (1978).
161. R. J. Knights, in *Nutrition for Special Needs in Infancy* (F. Lifshitz, ed.), Marcel Dekker, New York, 1985, pp. 105–115.
162. A. Voller, D. Bidwell, and A. Bartlett, in *Manual of Clinical Immunology*, 2d ed. (N. R. Rose and H. Friedman, eds.), American Society for Microbiology, Washington, D. C., 1980, pp. 359–371.
163. E. Young, M. D. Stonoham, A. Petruckevitch, J. Barton, and R. Rona, *Lancet 343*(7): 1127 (1994).
164. M. B. Ratner, S. Dworethky, and L. Aschim, *Pediatrics 22*: 648 (1958).
165. T. Kaneko, B. T. Wu, and S. Nakaik, *J. Food Sci.* 50(6): 1531 (1985).
166. J. J. Boza, J. Jiménez, O. Martinez, M. D. Suárez, and A. Gil, *J. Nutr. 124*: 1978 (1994).
167. H. L. Leary, *J. Pediatr. 121*: S42 (1992).
168. S. L. Bahna, *Ann. Allergy. 41*: 1 (1978).
169. T. Matsuda and R. Nakamura, *Trends Food Sci. Tech. 4*(9): 289 (1993).
170. D. P. Cuthbertson, *J. Sci. Food Agric. 1*: 35 (1950).
171. H. Otani, X. Y. Dong, and A. Hosono, *Milchwissenschaft 45*: 217 (1990).
172. C. T. Cordle, M. I. Mahmoud, and V. Moore, *J. Pediatr. Gastroenterol. Nutr. 13*: 270 (1991).
173. S. K. Kim, P. S. W. Park, and K. C. Rhee, *J. Agric. Food Chem. 38*: 651 (1990).
174. M. L. A. Casella and J. R. Whitaker, *J. Food Biochem. 14*: 453 (1990).
175. J. R. Whitaker and A. J. Puigserver, *Adv. Chem. Ser. 198*: 57 (1982).

176. T. Nakamura, H. Sado, and Y. Syukunobe, *Nippon Shokuhin Kogyo Gakkaishi 39*(1): 113 (1992).

177. T. Nakamura, H. Sado, and Y. Syukunobe, *Anim. Sci. Technol.* (Japan) *63*(8): 814.

178. G. Sawatzki and G. Georgi, in *Recent Advances in Infant Feeding* (B. Koletzko, A. Okken, J. Rey, B. Salle, and J. P. Van Biervliet, eds.), Georg Thieme Verlag, 1992, pp. 80–86.

179. J. J. Pahud, J. C. Monti, and R. Jost, *J. Pediatr. Gastroenterol. Nutr. 4*: 408 (1985).

180. T. Nakamura, H. Sado, Y. Syukunobe, and T. Hirota, *Milchwissenschaft 48*(12): 667 (1993).

181. J. Asselin, J. Amiot, S. F. Gauthier, W. Mourad, and J. Hebert, *J. Food Sci. 53*: 1208 (1988).

182. J. Asselin, J. Hebert, and J. Amiot, *J. Food Sci. 54*: 1037 (1989).

183. R. Jost, J. C. Monti, and J. J. Pahud, *Food Technol. 41*: 118 (1987).

184. J. J. Sullivan, and G. R. Jago, *Aust. J. Dairy Technol. 27*: 98 (1972).

185. J. Adler-Nissen, *J. Chem. Tech. Biotechnol. 34B*: 215 (1984).

186. M. Watanabe, T. Suzuki, Z. Ikezawa, and S. Arai, *Biosci. Biotech. Biochem. 58*(2): 388 (1994).

187. K. Miyakawa, Y. Hirai, J. Miyakawa, T. Sugiyama, T. Komatsu, S. Suga, Z. Ikezawa, and H. Nakajima, *Jp. J. Allergol. 37*: 1101 (1988).

188. K. Yamada, M. Kishimoto, Y. Inagaki, M. Inamoto, H. Komada, M. Yamada, and S. Torii, *Jp. J. Pediatrics 38*: 2545 (1985).

189. B. F. J. Goodwin and P. M. Rawcliffe, *Food Chem. 11*: 321 (1983).

190. M. Polgár, É. Gelencsér, and Gy. Hajós, *Élelmezési Ipar 47*(7): 204 (1993).

191. M. Polgár, thesis, 1993.

192. H. Izumi, T. Adachi, N. Fujii, T. Matsuda, R. Nakamura, K. Tanaka, A. Uritsu, and Y. Kurosawa, *FEBS 302*: 213 (1992).

193. M. Watanabe, J. Miyakawa, Z. Ikezawa, Y. Suzuki, T. Hirao, T. Yoshizawa, and S. Arai, *J. Food Sci. 55*(3): 781 (1990).

194. M. I. Mahmoud, W. T. Malone, and C. T. Cordle, *J. Food Sci. 57*(5): 1223 (1992).

195. Gy. Hajós, É. Gelencsér, Á. Pusztai, G. Grant, M. Sakhri, and S. Bardocz, *J. Agric. Food Chem. 43*: 165 (1995).

196. P. Pellett and V. R. Young, in *Nutritional Evaluation of Food Protein*, United Nations University, 1980.

197. A. Pusztai, in: *Plant Lectins, Chemistry and Pharmacology of Natural Products*, Cambridge Univ. Press, Cambridge, 1991.

198. Gy. Hajós, É. Gelencsér, S. Bardocz, A. Pusztai, and G. Grant, unpublished, 1995.
199. M. Hessing, H. Bleeker, M. van Biert, H. A. A. Vlooswijk, and R. J. Hamer, *Med. Fac. Landbouww Univ. Gent* 59(4b): 2257 (1994).
200. M. Tanaka, K. Csúri, and J. Molnár, unpublished results, 1994.
201. J. Salji, *Food Sci. Tech. Today* 8(3): 139 (1994).
202. A. Shimada, I. Yamanoto, H. Sase, Y. Yamazaki, M. Watanabe, and S. Arai, *Agric. Biol. Chem.* 48: 2681 (1984).
203. É. Gelencsér, M. Polgár, and Gy. Hajós, Abstract of the First IUBMB Conference, Nagoya, Japan, 3a 06-9.

7
Denaturation of Globular Proteins in Relation to Their Functional Properties

Jacques Lefebvre

Institut National de la Recherche Agronomique, Nantes, France

Perla Relkin

Ecole Nationale Supérieure des Industries Alimentaires, Massy, France

I. INTRODUCTION

Because of the tightly packed character of their native structure, entropy tends to unfold the native conformation of globular proteins and must be compensated for by the interplay of intraprotein and protein–solvent noncovalent interactions.

The thermodynamic stability of the native conformation arises from the minimization of the overall free energy of interaction. It results from a unique balance between large stabilizing and large destabilizing forces and appears to be only marginal.

Therefore, relatively limited changes in the external variables of the system protein + solvent are able to destabilize the structure of the protein, i.e., to induce its unfolding. This is usually referred to as protein denaturation. The expression, however, is ambiguous since it is frequently used to designate in fact the conjunction of two distinct events:

—The process of unfolding, which is in principle reversible, as recognized very early in the history of the physical chemistry of proteins and firmly established by Anfinsen more than 30 years ago (see for example Ref. 1).

—Irreversible changes in the system, concomitant to unfolding or a consequence of it. They can be, in extreme conditions, chemical reactions modifying some amino acid residues, with the possible formation of covalent intra- and intermolecular cross-links; the most usual cases are SH group oxidation, disulfide bond reduction, and sulfhydryl/disulfide interchange reactions. But generally they result from noncovalent aggregation of the unfolded molecules. Aggregation triggered by unfolding competes with refolding in most types of denaturation, as soon as protein concentration is not very low.

The study of protein denaturation is relevant to food science and technology in different respects.

—Biological functions (enzymatic activities, specific ligand binding, etc.) are intimately governed by the native conformation and are lost upon denaturation, and so those of the functional properties, in the technological sense of the terms, which derive directly from molecular characteristics of the native structure. This aspect of the question is the best known, but not necessarily the most important, as to protein "functionality."

—Denaturation underlies functional properties of globular proteins as texture agents. Except in very specific and rare cases of strong intermolecular interactions, native globular proteins, because of their corpuscular nature, have no appreciable thickening or gelling properties. These properties develop actually as a consequence of denaturation (usually thermal) and derive from the strong tendency of unfolded proteins to aggregate in an aqueous medium. On the other hand, a particular type of denaturation is involved in foaming and emulsifying properties of globular proteins; some degree of unfolding occurs upon adsorption of the protein at the interface and can result in interfacial aggregation or gelation of the protein, both steps being important as to the considered properties.

In spite of a tremendous amount of work, several very important aspects of protein denaturation in solution remain ill-known or controversial, because of theoretical and experimental difficulties. These difficulties are still much more acute in the case of the much less studied interfacial denaturation of proteins; interfacial denaturation follows the same general lines as solution denaturation, but with the additional constraints of taking place in a thermodynamically aniso-tropic and spatially confined (nearly two-dimensional in its initial stages) environment.

Protein denaturation can be caused by a large number of physical and chemical factors (for an overview, see for example Refs. 1–3). We shall focus here on thermal denaturation (in solution), since it is obviously of prime importance in food science and technology, and also for fundamental reasons of its direct links with the thermody-namics of protein unfolding. Interfacial denaturation will be treated more succinctly, for less information is available on it.

Many of the most important food proteins are not globular. Fibril-lar proteins, of particular concern for meat science and industry, are also rigid particles with internal organization and present the same types of secondary structures as globular proteins, but they display extended overall shapes; they share basically the same denaturation properties with globular proteins. Prolamins and caseins constitute quite different types of proteins; since they are much less compact and much less organized at the molecular level than globular pro-teins, the concept of denaturation cannot apply fully to them. We shall restrict ourselves in this chapter to globular proteins, which are the most typical and classical proteins; as often as possible, we shall take as actual examples bovine serum albumin, β-lactoglobulin, and ovalbumin, which are of interest in food science and technology.

II. THE NATIVE CONFORMATION

Study of protein denaturation, i.e., the changes in protein conforma-tion under the effect of different factors, requires evidently some knowledge of both the native and the denatured conformations. Al-though the precise native structure of globular proteins at the atomic scale is in general not available, a fairly good characterization of the native conformation is usually achieved straightforwardly if labori-ously. However, some problems pertaining to the definition of the native state can arise in relation with denaturation studies but are often overlooked.

Among the proteins we have chosen as examples, the crystallo-graphic structure of bovine β-lactoglobulin (BLG) was the first determined. The crystal structure of ovalbumin has been reported only recently [4]. Bovine serum albumin (BSA) can be readily crystallized but, probably because of a rather flexible and fluctuating conformation, crystallographic resolution is too poor to give precise information on its structure. Nevertheless, the structure at 2.8 Å resolution of human serum albumin (HSA), a protein very close to BSA, has been published [5,6]. In most cases, structural information comes from indirect or relative methods, which can be also used to follow conformation changes. Description of the methods used to study protein conformation, of their application to the selected proteins, and analysis of the results thus obtained are utterly out of our scope. The main conformational characteristics of our proteins are surveyed in the recent monographs by Peters [7a] and by Carter and Ho [7b] on BSA, and by Hambling et al. [8] on BLG, to which the reader is referred. No such recent, extensive, and thorough review exists in the case of ovalbumin, for which moreover structural information is less abundant; for lack of an up-to-date account, one can find still useful information in the paper of Taborsky [9].

The three proteins chosen are good examples to bring to light some of the practical difficulties, important in view of denaturation studies, encountered when one tries to define the native state of a protein.

BSA, ovalbumin, and BLG share the characteristic of possessing both disulfide and thiol groups. They are therefore prone to SH/S-S interchange and intermolecular disulfide bond formation at neutral or basic pH values, especially in denaturing conditions. This problem will be considered later in relation to denaturation. But it can cause already some heterogeneity in the "native" protein that has been submitted to conditions inducing appreciable SH and S-S accessibility without denaturation. As is well known, gel electrophoresis of commercial BSA samples frequently reveals faint slow-moving bands that correspond probably to "polymers" of the protein bound by intermolecular disulfide linkages formed during its purification. This heterogeneity is actually an artifact.

BSA is able to bind a great variety of ligands, and this results in microheterogeneity. One of its biological functions is to transport fatty acids; its native molecules bear therefore different amounts of bound fatty acids that affect their properties; for example, the isoelectric point of defatted BSA is 5.2, instead of ~4.8 for the nondefatted protein, and the shape of the nondefatted molecule differs slightly from that of the defatted one [10]. Moreover, nondefatted molecules are more resistant to heat denaturation and less prone to subsequent aggregation than defatted ones.

Ovalbumin is a phosphorylated and glycosylated protein with different degrees of phosphorylation [9] and different carbohydrate moieties [11], and consequently it shows a large microheterogeneity. More important certainly in a practical perspective is the existence of a particular form of the protein, called S-ovalbumin, the proportion of which is small relative to total ovalbumin in fresh egg but increases enormously during storage [12,13]. Although conformation and most properties of both forms seem to be very similar, S-ovalbumin is more stable to heat denaturation [13].

BLG illustrates other types of heterogeneity in the native state that can complicate the study of its denaturation. BLG exhibits complex multimerization equilibria and conformational polymorphism, extensively studied, especially by Timasheff and coworkers, and summarized in [8] and [14], which depend on pH, ionic strength, and protein concentration, but also on temperature [15–17]. Besides, BLG shows genetic polymorphism; the main genetic variants are A and B, which are present in about equal quantities in most samples. Though the sequences of the A and B variants differ only at positions 64 and 118 [8], and even though their native secondary structures appear to be nearly identical [18,19], this slight difference suffices to induce noticeable changes in some of their properties, highlighting how tight the relation can be between the structure and the properties of proteins. Compared to BLG-B, BLG-A is reported to show higher self-association proclivity between pH 3.7 and 5.1 (dimer to octamer transition) [14] but a lower one at pH 2.7 (monomer to dimer transition) [20], lower apparent enthalpy of thermal denaturation [21,22] and firmer heat-set gels [23] at neutral pH in presence of 0.1 M salt. A more insidious difficulty arises from the multimerization equilibria of BLG. Except in extreme conditions where BLG is in the monomeric but still native form (for example at pH 2 and low ionic strength), solutions of the protein will contain both monomers and dimers, or dimers and octamers, in proportions depending on the conditions. Therefore characteristics such as those pertaining to the size and shape will be determined generally as apparent ones and their modifications due to different factors could reflect displacement of the multimerization equilibria as well as conformational changes.

Some aspects of protein microheterogeneity in relation to their denaturation behavior will be examined in more detail later on.

Finally, it turns out that different globular proteins have some tendency to aggregate in aqueous solutions as soon as their concentration is not very low. For example, Matsumoto et al. observed in small angle x-ray scattering (SAXS) the existence of aggregates of ovalbumin resulting in an average state of association number of nearly 2 in 0.5–1% solutions at pH 7, whereas the protein was dispersed as individual molecules at concentrations below 0.1% [24,25]. They determined in a similar way that BSA has an average degree of association of ~1.8 in the concentration range 0.1–1% at pH 7 in 20 mM sodium phosphate buffer [26]. Previously, glutamate dehydrogenase and β-galactosidase were shown to associate (in a linear and in a random fashion, respectively) in buffer solutions as their concentration increases [27].

Proteins in solution are also sensitive to orthokinetic aggregation, i.e., shear-induced aggregation. We have observed this phenemenon at very low shear rates in the case of BSA, ovalbumin, and BLG [28,29]; in Fig. 1, one sees that the viscosity of the protein solution increases with time till it reaches a plateau; the phenomenon becomes less and less pronounced as the shear rate increases, so much so that the solution displays time-independent and Newtonian flow behavior in the usual shear rate range.

III. THE THERMODYNAMICAL APPROACH TO PROTEIN UNFOLDING AND STABILITY: AN OVERVIEW

Unfolding of the native conformation is the primary process in protein denaturation. It consists in an extensive but reversible loss of the original three-dimensional organization of the macromolecule. Since

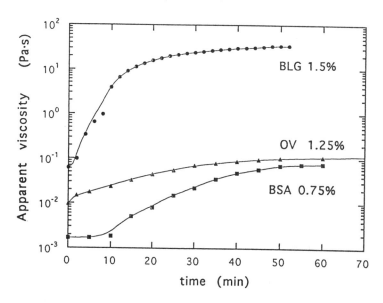

Figure 1 Shear-induced aggregation of bovine serum albumin (BSA), ovalbumin (OV) and β-lactoglobulin (BLG) in aqueous solution. Viscosity of the solutions was measured against time under a steady shear rate (10^{-3} s^{-1} for BSA, $6.3 \ 10^{-3} \ s^{-1}$ for OV, and $10^{-4} \ s^{-1}$ for BLG). Solvent: 0.1 M NaCl, pH 7, 20°C. The concentrations of the solutions were 0.75%, 1.25%, and 1.5% for BSA, OV, and BLG, respectively. (Replotted from Refs. 28 and 29.)

mechanisms of unfolding are intimately related to protein structure and stability, their study is of utmost interest for the understanding of protein folding and structure–function relationships. Accordingly, protein unfolding has been the subject of an enormous amount of work, the different aspects of which it is out of our purpose to sum-marize,* even in the more delimited field of thermal unfolding to which we restrict ourselves. The following short account will focus on the question from the thermodynamical standpoint, i.e., on its equilibrium aspect, kinetic aspects and structural pathways of un-folding being left aside.

---

*For general reviews, see for example [1,2,3,30,31,32].

A. Folding/Unfolding Transition

The equilibrium thermodynamical approach implies the unfolding process to be reversible. Reversibility is a consequence of native structure being a minimum energy conformation, and it is in close relation to the highly cooperative character of the transition from the native to the completely unfolded state [33–35]. Cooperativity is reflected in the negligible amount of stable intermediate states between the native and unfolded conformations (two-state behavior), or in the small number of these states as compared to the very large number of possible intermediates (multistate transitions) [33].

The simplest mechanism of the folding/unfolding transition is the two-state one in which the only conformations populated with a significant fraction of molecules at any point during the transition are the native state (N) and the unfolded one (U). Small proteins (Mr < 20,000) usually conform to this mechanism [35]; that means that they behave as one single cooperative unit, as we have already pointed out. Such a mechanism corresponds to the equilibrium $N \rightleftarrows U$, with the equilibrium constant $K = [U]/[N]$ being in the range 10^{-11}–10^{-3} in "native" conditions: even in optimal conditions for stability of the native conformation the protein has a finite probability of being in the unfolded state, an illustration of the marginal character of protein stability. A change in an external variable promoting the denaturation, for example an increase in temperature T, will increase K considerably, shifting the equilibrium to the right. The thermodynamical functions associated with the transition are derived straightforwardly from $K(T)$:

$$\Delta G_N^U(T) = \Delta H_N^U(T) - T\Delta S_N^U(T) = -RT \ln K(T)$$

$\Delta G_N^U(T)$, $\Delta H_N^U(T)$, and $\Delta S_N^U(T)$ being the variations of Gibbs free energy, of enthalpy, and of entropy, respectively, upon unfolding.

Larger proteins are generally multidomain, and their unfolding transition passes often (but not always) through a small number of partially folded intermediates (X_i) that are present in significant amounts, corresponding usually to the more or less independent unfolding of the different structural domains, which behave as cooperative folding units [36]. The transition will be then represented by a series of equilibria between n states:

$$X_1 \rightleftharpoons X_2 \rightleftharpoons \cdots \rightleftharpoons X_{n-1} \rightleftharpoons X_n \qquad (X_1 = N; \ X_n = U)$$

Its study would require the identification of the different states and the determination of the Gibbs free energy differences relative to each step. The thermodynamical analysis of multistate transitions is especially complex if the domain transitions are not independent (i.e., in case of domain–domain interactions). A general formal treatment is given in [33,34]; a simple model for the particular case of a protein with two interacting domains is detailed in [37].

B. Experimental Approach

1. Indirect Methods

Almost any method able to monitor conformational changes has been used to study protein unfolding. Of particular importance have been techniques that are very sensitive (allowing work at very low concentration and limiting therefore the problems resulting from aggregation of the unfolded form) and give quantitative information on the secondary structure (optical rotation dispersion, far-UV circular dichroism, Raman spectra analysis) or on the environment of specific residues (solvent perturbation techniques and differential UV spectroscopy, fluorimetry, near-UV circular dichroism, NMR), or on the global accessibility of the amino acid residues to the solvent (hydrogen/deuterium exchange).* In the case of a two-state transition, they allow us to measure, in relative terms at least, the concentration of the two species. If the measured parameter at a given value of the external variable considered is ν, and its values relative to the native and unfolded state are ν_N and ν_U, respectively, ν_N and ν_U being assumed proportional to the concentrations of N and U, then the degree of unfolding is $x = (\nu - \nu_N)/(\nu_U - \nu_N)$, and the equilibrium constant is expressed as $K = x/(1 - x)$.

However, this approach can be seriously misleading when the transition cannot be any longer approximated by a two-state one [38]. Departure from the two-state approximation can be detected only through complicated kinetics studies leading to results difficult to analyze [1,30,32].

---

*For reviews on the application of different methods see [1,2,3,31,32].

2. Differential Scanning Calorimetry

The limitation of calorimetric methods for the study of protein unfolding has been for long the lack of sensitivity, which compelled us to carry out the measurements at rather high protein concentrations. Then, what is monitored is usually the global denaturation process, i.e., as we have already mentioned, the conjunction of protein unfolding, a reversible endothermic process, and aggregation of unfolded molecules, an irreversible and exothermic one. This hinders thermodynamical analysis, since it requires reversibility of the process, unless the overlap of the two processes is sufficiently small or the aggregation is kept down to a negligible level in the conditions of the measurements, conditions that are not frequently met.

Since the beginning of the 1970s, high-sensitivity and precise microcalorimetry, and particularly differential scanning calorimetry (DSC) techniques, have been developed as powerful tools for the study of protein unfolding. An especially important contribution to this development has been that of Privalov and coworkers. High-sensitivity micro-DSC operates at the 0.1% concentration order of magnitude, so that aggregation has a negligible incidence on the results.

The methodology of high-sensitivity DSC and application of the technique to the study of heat-induced changes in proteins have been expounded in several reviews [35,36,39–44], to which the reader can refer. The following is just to outline its main features in order to facilitate subsequent discussion.

DSC allows the determination of the thermodynamic functions of the unfolding transition directly from the heat capacity curve recorded. The analysis requires only the transition to be reversible, which can be easily checked, as will be pointed out below; no a priori assumption on the number of states or on the mechanism of the transition is needed.

The heat capacity peak, characteristic of the transition, reflects the excess heat capacity arising from the enhanced enthalpy fluctuations that occur in the temperature range of the transition. In the case of a two-state transition, the thermodynamic functions are obtained in a straightforward way from the area Q_D under the peak (corrected for the baseline), which measures the overall enthalpy change resulting from the transition, and the overall heat capacity difference $(\Delta C_p)_N^U$

between the unfolded and the native states, both referred to the transition temperature T_0, defined as the temperature at mid-transition corresponding to [U] = [N]. T_0 can be approximated by the temperature T_m of the maximum of the recorded DSC peak; the calorimetric overall enthalpy $\Delta H_N^U(T_0) \equiv Q_D$ is identical to the van't Hoff enthalpy ΔH_{vH} of the process, which can be calculated from $(\Delta C_p)_N^U$, T_m, and the height of the peak [40].

In the case of a multistate transition, the partition function (the sum of the statistical weights of all the states) can be calculated from DSC data, since the population of each state is a function of its enthalpy, and deconvolution of the DSC peak(s) leads to the number of the states and to the determination of the thermodynamical parameters for each of them [39,40,43,44]. Other experimental methods (cf. Sec. III.B.1) do not offer this possibility, since the variables they measure are not related to a thermodynamic function. However, deconvolution of DSC data is not an easy task, especially when the different steps of the transition strongly overlap, with the result of a single DSC peak the shape of which is hardly distinguishable from that of a two-state transition; besides, whereas the overall thermodynamical parameters ΔH_N^U and ΔS_N^U are not very sensitive to the method of baseline correction [45], baseline determination is critical for deconvolution [44]. It can be interesting to decrease the overlap of the transition steps by changing solvent conditions [40]. In the method called "two-dimensional DSC," DSC scans are performed at systematically varying concentrations X of a denaturant or of a ligand affecting thermal stability, and the thermal stability surfaces thus generated in the (C_p, T, X) space are analyzed in terms of the linkage equations between the variables T and X [44,46–48].

The ratio $r = Q_D/\Delta H_{vH}$ is frequently used as a touchstone to check in a simple way whether the transition departs appreciably from a two-state one [35,40,49]. A ratio significantly higher than 1 would point to a multistate process, even though only one calorimetric peak were observed. However, Freire [43,44] has recently drawn attention to the fact that r does not provide an accurate assessment of the character of the transition and actually can be misleading [43,44,48]. A ratio $r < 1$ is considered as indicative of an irreversible process, generally due to aggregation of the unfolded species [40].

Irreversibility prevents the equilibrium thermodynamical analysis of a process but not the characterization of its energetics by DSC; in this case the area Q_D below the peak represents the heat of reaction. So, DSC can be used to study the overall process of denaturation. Checks for reversibility and aggregation effects in DSC will be illustrated later (Sec. VI).

The fact that both unfolding and aggregation contribute to DSC response could be turned to advantage in the perspective of this chapter. It should be possible to extract the contribution of aggregation in order to study the energetics of this process, as far as the contribution of unfolding can be evaluated separately, using high sensitivity DSC and very low protein concentration for example. It appears that this possibility has not been exploited.

Finally, DSC permits the determination of the absolute values of heat capacity in the native and the unfolded states. These values are related to the amino acid residues exposed to water in each state and therefore contain structural information on these states. This provides the basis for an analysis of the thermodynamics of unfolding in terms of structure [43,44], some aspects of which will be sketched below.

C. Folding/Unfolding Thermodynamics and Protein Stability

The results of high-sensitivity DSC studies of protein denaturation have greatly helped to clarify the reversibility and the intermediate states issues, as indicated above. They have yielded a detailed analysis of the thermodynamical features of protein unfolding and led to a reassessment of the contributions of the different forces that determine protein stability.

1. The Main Thermodynamical Characteristics of Protein Unfolding

The experimental results on protein unfolding thermodynamics have been extensively reviewed [35,36,41,50,51]. We have already considered the reversibility and the two-state assumptions; let us now examine briefly the thermodynamical functions as they are determined by the application of DSC.

Upon unfolding, an increase $(\Delta C_p)_N^U$ in the specific heat capacity of the protein, amounting to 5–25 kJ K^{-1} mol^{-1} at 25°C, is measured. The differences in $(\Delta C_p)_N^U$ between proteins are due to differences in the specific heat capacities of their native states; the specific heat capacities of the unfolded states are similar and do not depend appreciably on the solvent conditions or the denaturing agent.

The existence of a significant $(\Delta C_p)_N^U$ entails that ΔH_N^U and ΔS_N^U are temperature dependent. They are actually rather complicated functions of temperature [52], since $(\Delta C_p)_N^U$ varies with temperature following a squat bell-shaped curve with a maximum usually in the 10–50°C range [41]. In particular, $\Delta H_N^U(T)$ is not linear, as shown experimentally in a recent paper [53]. Their values in the 20–80°C range differ little from those calculated with the assumption of a constant $(\Delta C_p)_N^U$ [41]; but deviations become important at lower temperatures [54].

A very remarkable feature of $\Delta H_N^U(T)$ and $\Delta S_N^U(T)$ is that these functions reach an upper limit above a "convergence temperature" $T^* \sim 140°C$, and that their specific limiting values $(\Delta H_N^U)^*$ and $(\Delta S_N^U)^*$ per gram or per amino acid residue are nearly the same for all "typical" globular proteins [41]. More recent results indicate that the convergence temperature is not the same for $\Delta H_N^U(T)$ (T_H^*) as for $\Delta_N^S(T)$ (T_S^*) [34].

As a consequence of the temperature dependence of ΔH_N^U and ΔS_N^U, $\Delta G_N^U(T)$ shows a bell-shaped curve; an example is given in Fig. 2. This has two important implications: (1) $\Delta G_N^U(T)$, which measures the stability of the protein, passes through a maximum at a temperature T_{max}, usually in the 0–50°C range, at which the native conformation is stabilized only by the enthalpy difference between the two states. As already stressed in the introduction, $\Delta G_N^U(T_{max})$ is rather small, albeit it results from comparatively large enthalpic and entropic effects; typical values are of the order of 50 kJ per mole of protein. (2) On both sides of T_{max}, stability decreases and there are two temperatures at which ΔG_N^U becomes zero, i.e., denaturation occurs: the upper one corresponds to the temperature of the usual heat denaturation process, the lower one to the somewhat paradoxical process of "cold denaturation"; the paradox lies in that cold denaturation is an exother-

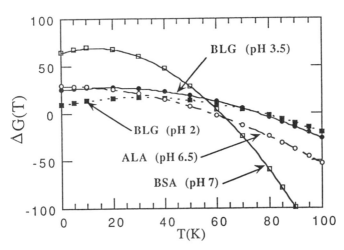

Figure 2 Examples of free energy change variation as a function of temperature for some examples of globular proteins. (Thermodynamic data obtained by DSC.) ALA (pH 6.5): α-lactalbumin at pH 6.5. (Data from Ref. 190.) BLG (pH 2): β-lactoglobulin at pH 2. (Data from Ref. 28.) BLG (pH 3.5): β-lactoglobulin at pH 3.5. (Data from Ref. 45.) BSA (pH 7): bovin serum albumin. (Data from Ref. 64.)

mic process driven by a decrease in enthalpy and entropy, which would be expected to favor on the contrary an ordered conformation. Cold denaturation of proteins, predicted initially by Brandts [55] and reviewed by Privalov [56], has been effectively observed in many cases, and experimentally studied by calorimetry for several proteins. It is generally a process more complex than heat denaturation, as shown, for instance, in the case of BLG [57] or of phosphoglycerate kinase [58,59].

As pointed out by Privalov [35], the stabilization energy value per residue is an order less than thermal energy kT, making evident that cooperativity plays a most important role in stability of the native conformation, since it integrates somehow the contributions of the elements constituting the protein. Cooperativity is reflected in the negligible number of stable intermediate states between the native and the unfolded conformations (two-state behavior), or in the small number of these states as compared to the very large number of possible intermediates (multistate transitions) [33].

Cooperativity of the folding/unfolding transition can be due to cooperativity of the interactions involved in the stability of the native state (i.e., the total contribution of the interactions is larger than their individual ones) or to unfavorable interactions in partially unfolded states. Both effects are indeed implied. Mutational studies have demonstrated for example synergies between the salt bridges of barnase [60,61], and cooperativity between the two disulfide bonds of ribonuclease T1 [62]. On the other hand, Freire and coworkers [33,34,63] have shown that what they call "local intermediates," i.e., intermediates in which a region of the protein is unfolded whereas the remainder keeps its native conformation intact, are highly unstable in the case of single-domain globular proteins; in such intermediates, in addition to the residues belonging to the unfolded parts of the molecule, nonpolar residues located at the interface of the regions undergoing unfolding would become exposed to the solvent, and the unfavorable free energy resulting from this additional solvent exposure is not compensated for by a significant conformational entropy gain.

As to multidomain proteins, they behave as single cooperative units if the interactions between the domains are about as strong as interactions within the domains; but the domains unfold independently if their interactions are weak, and the partially folded intermediates correspond to conformations in which one domain is folded and the others unfolded [36]. Complex unfolding behavior with more than one endothermal peak can be in some instances observed in DSC, as for instance phosphoglycerate kinase [59]. Although BSA comprises three domains, it shows generally only one peak upon heat denaturation and is even reputed to conform apparently to the two-state approximation [35]; anomalies in the values $(\Delta H_N^U)^*$ and $(\Delta S_N^U)^*$ were however spotted [35], and we observed recently that the ratio r indicates a strong departure from the two-state approximation [64]. Nonetheless, in an early study [65] Privalov and Monaselidze reported a thermogram with three denaturation peaks (at 57, 69 and 77.5°C, respectively) upon heating at a very slow rate (0.06°C min^{-1}) for a 2% BSA solution in pH 7 phosphate buffer. Aoki et al. [66,67] obtained, for BSA with its -SH group blocked, two endothermal peaks in the pH range 4.3–7.3 in NaCl 0.01 M, and in the pH range 4.2–5.2 only when the salt concentration was 0.1 M. They interpreted their results as indicating an independent unfolding of two thermodynamical domains of BSA in moderately acidic to neutral conditions, and a stabilizing effect of NaCl by screening electrostatic repulsions that tend to open a cleft between the domains [66,67]. This interpretation seems reasonable; the disappearance of the high-temperature DSC peak at pH < ~4.5 could reflect the "acidic expansion" of BSA that occurs below pH 4.5. However the utmost prudence should prevail in analyzing this type of result since aggregation certainly occurs in the conditions of their DSC measurements, and its importance will depend sharply on the pH and on the ionic strength.

Some proteins can show either two-state or multistate behavior depending on the conditions of denaturation. For example, heat denaturation of phosphoglycerate kinase complies with the two-state approximation, but not its cold denaturation [58]; this has been quantitatively analyzed as resulting from the existence of two domains, interacting mainly through hydrophobic contacts, that unfold almost independently at low temperature but behave as a single cooperative unit at high temperature, as a consequence of the temperature dependence of hydrophobic interactions [59,63].

2. Interactions Determining Protein Stability

Stability of the native structure represents the balance of the noncovalent interactions between the chemical groups of the protein and the interactions between these groups and the surrounding water molecules. The thermodynamical study of protein unfolding gives just the balance sheet of the energy terms relative to the different noncovalent interactions responsible for protein stability. Assessment of their respective contributions based on the analysis of calorimetric data has been the subject of much effort.

For the purpose of the analysis of the energetics of protein unfolding, the temperature dependence of the thermodynamical parameters of the transfer to water of various low-molecular-weight compounds has been used to estimate hydration effects, applying the group additivity method in conjunction with the determination of water-accessible surface area changes upon unfolding of proteins of known tridimensional structures [34,41,54,68–73]. Attention has been focused particularly on the interpretation of the large heat capacity increase [34,54,68,73] and of the convergence temperatures T_H* and T_S* [34,41,69,72] associated with protein unfolding, features that are both also displayed upon transfer to water of nonpolar small molecules.

The correlation of $(\Delta C_p)_N^U$ with the buried nonpolar surface area or the number of van der Waals contacts between nonpolar groups in the native structure was interpreted until recently as an evidence of the determinant contribution of nonpolar groups' exposure to protein unfolding energetics, and therefore of hydrophobic interactions to protein stability [35,49,74]. This interpretation conformed to the current idea that hydrophobic interactions* are the driving force in protein folding and constitute by far the main energetic contribution to protein stability, which has been for long prevailing. This view amounts in substance to the belief that protein folding results from the entropically driven reduction in nonpolar surface area accessible to water in the folded conformation as compared to the unfolded state; the entropy gain deriving from the shielding of apolar groups from water would overcompensate for the entropy loss upon folding, hence the stability of the compact native structure. It still has distinguished supporters [76–78] and has even been tentatively regrounded by transposition of the macroscopic concept of interfacial free energy to the molecular level [78,79]; for a short discussion of this approach see also [80]. However, the hydrophobic effect on its own cannot explain such a basic fact as the conformational specificity of native proteins, although it does explain the collapse of the polypeptide chain into a compact structure [75,81]; it seems moreover difficult to understand thermal unfolding of proteins on the sole basis of hydrophobic interactions, which are reputed to strengthen as temperature increases at least up to 60°C.

Since the mid-1980s, the importance of hydrophobic interactions to protein structure has been much disputed [75]; as happens often in such instances, part of the confusion is only apparent and originates

---

*For short and relatively recent accounts of the hydrophobic effect and its relation to protein folding and stability, the reader can refer to [75] and [76].

in semantic divergences (cf. Refs. 70, 82, 83). A much more balanced picture has progressively emerged.

In fact, unfolding of the protein native structure exposes to the solvent not only nonpolar groups but also a large number of polar groups (about 40% of the total buried surface area is polar), mainly peptide bonds. Hydration of polar groups does contribute to $(\Delta C_p)_N^U$; this contribution is negative, but its magnitude decreases as temperature increases, whereas that of the apolar groups is positive and decreases as temperature increases [68]. A third contribution to $(\Delta C_p)_N^U$ is expected: that resulting from the increase of the conformational freedom upon unfolding; it seems however that it is nearly offset by another effect, probably linked to the fluctuations of the native structure, which increase as temperature increases [68]. The convergence of the specific values of $\Delta H_N^U(T)$ and $\Delta S_N^U(T)$ to $(\Delta H_N^U)^*$ and $(\Delta S_N^U)^*$ at temperatures T_H^* and T_S^* is interpreted as resulting from the combination of a polar contribution to the thermodynamic quantities that is nearly the same for all "regular" globular proteins, reflecting a nearly identical number of hydrogen bonds (0.72) per residue, with a nonpolar contribution that varies from one protein to another [34].

The analyses of Privalov and Makhatadze [69–71] and of Murphy and Freire [34,72] converge in pointing up the importance of the enthalpy change linked with the transfer of polar groups from the close-packed interior of the native structure to water. This change includes both the disruption of intramolecular hydrogen bonds between the buried polar groups and the hydration of these groups. Hydration of polar (and aromatic) groups tends to destabilize the native state, as does the conformational entropy change; the first effect, which results in a negative free energy contribution the magnitude of which increases as temperature decreases, causes cold denaturation, and the second effect, heat denaturation [70,71]. Although there is disagreement on quantitative grounds, especially about the calculation of the entropic terms [72,84], a conclusion that can be drawn from these studies, substantiated by further recent DSC experimental results on the unfolding of different proteins [53,85–87], is that intramolecular hydrogen bonding is, together with van der Waals interactions, a major contributor to the stability of protein native conformation.

The importance of hydrogen bonding in protein folding and stability is also stressed by other approaches such as computer simulations [80] and mutational studies [88], while Ben-Naim has shown the fallacy of the argument of a nearly zero contribution of hydrogen bonding to ΔG_N^U [38]. Formation of intramolecular hydrogen bonds in polypeptides appears to be energetically preferential to the hydration of their donors and acceptors [34,80]. Moreover, all intramolecular hydrogen bonds do not seem to be energetically equivalent; it appears that upon heat denaturation a hydrogen bond is stronger in a β-sheet structure than in an α-helix [87], a fact that could explain the predominance of β-sheets in the residual structures observed in some instances in heat denatured proteins (see below).

Intermolecular disulfide bonds contribute of course to protein stability. For example, it has been demonstrated that ribonuclease A with its four disulfide bonds reduced unfolds even in the absence of a denaturing agent [32]. The stabilizing effect of disulfide cross-links is very significant (up to 16 kJ mol^{-1} per disulfide bridge) as demonstrated by mutational and protein engineering studies reviewed in [88,89], although introduction of "artificial" cross-links can prove highly detrimental to protein stability because they can introduce significant strain into the molecule. What is actually expected is a decrease in the entropy of the unfolded state [35,80], resulting in an enhanced stability of the native conformation. However, recent DSC studies bring to light that disulfide bridges affect also significantly the enthalpy of unfolding in favor of the native state [54]. On the other hand, the existence of a disulfide bridge causes a decrease in the accessible surface area of the unfolded state; this effect, which "stabilizes" the unfolded state and consequently is unfavorable to the stability of the native conformation, can offset the favorable configurational entropy effect [90]. An additional contribution to the stabilizing effect of disulfide bridges could come from their putting limitations to the fluctuations in the native conformation.

Electrostatic interactions play a minor role in protein stability [76,91], albeit they can be of importance to the specificity of protein structure or protein interactions [32].

IV. CONFORMATION OF THE DENATURED STATE(S)

In the analysis of the energetics of protein denaturation, knowledge of the structure of the unfolded state is as important as the knowledge of the native one. The thermodynamical approach to protein denaturation usually assumes that the result of denaturation is the complete exposure to the solvent of all the amino acid residues and peptide bonds, which means that the unfolded state is one single population of highly expanded, fully solvated and disordered conformations, whatever the protein, the denaturation process, or the solvent conditions considered.

The main argument put forward is the close similarity of thermodynamical characteristics of protein denaturation when achieved under the effect of different agents, which has been considered as precluding any significant thermodynamical contribution of a possible residual structure in the denatured state and as an evidence that the extent of unfolding is the same for all denaturational processes [35,49]. Since it had been demonstrated that proteins in 6 M guanidinium hydrochloride (GuCl) or 8 M urea adopt as a rule the random coil conformation when disulfide bonds are reduced (or the cross-linked random coil one when they are not) [2,3], it was a short step to infer that the final state of any denaturation process is a completely unfolded conformation, if not the random coil strictly speaking [35].

However, many experimental studies indicate that denatured globular proteins do not show necessarily complete exposure to solvent but can be more or less compact conformations or even can retain significant amounts of ordered structures; besides, there is experimental evidence that for a given protein different denatured states can exist, depending on the denaturation agent and conditions.

The degree of swelling of a random coil depends on the "quality" of the solvent. While 6 M GuCl or 8 M urea are "moderately good" solvents for randomly coiled proteins [2,3], water is certainly a rather bad one for denatured proteins, as demonstrated by the compact structure of the native state and suggested by their strong propensity to aggregate when unfolded in water, after heat denaturation for example. The problem is to know whether the size of denatured proteins in water is larger or smaller than the unperturbed value, corresponding to the minimum size for a randomly coiled conformation, reached in "θ conditions," which can be estimated in different ways [2,3,92–95], though approximately [2,92,96,97]. The latter case would correspond to a collapsed coil, i.e., a compact conformation.

The criterion however applies only to uncross-linked proteins, so that no conclusion can be drawn from the rather small increase, circa 20%, in the Stokes radius R_s upon heat denaturation measured by dynamic light scattering (DLS) for chymotrypsinogen, ribonuclease, and lysozyme [98], and for ribonuclease [99]; for comparison, 5 M GuCl causes a 45% increase to R_s of unreduced lysozyme [100].

An expanded conformation $(R_s > (R_s)_\theta)$ does not mean necessarily a randomly coiled one or complete exposure of amino acid residues: because of the heteropolymeric nature of the polypeptide chain, the solvent can be "good" for some parts of the sequence but "bad" for others, with the result that the chain can be locally collapsed but its overall conformation expanded. Therefore, nothing can be inferred from the large radii of gyration R_g measured by DLS or SAXS for apocytochrome c (ACC) [101] and β-lactamase I (BLM) [102] at low ionic strength once denatured at low and high pH, respectively, although ACC and BLM are uncross-linked proteins. When intramolecular charge repulsions are screened by shifting the pH to less extreme values or by adding salt, these radii drop considerably [101,102]; they become indeed lower than the estimated unperturbed radii $(R_g)_\theta$ of the equivalent randomly coiled chains [97], indicating that denatured ACC and BLM show compact conformations in such conditions. In the case of BLM, restoration of a substantial amount of ordered structure accompanies this shrinkage [102].

The radius of gyration of staphylococcal nuclease (SNase), measured by SAXS, increased from 1.6 nm in the native state to 3.3 nm in 8 M urea [103]. The latter value is however well below the one (4.3 nm) that is expected for $(R_g)_\theta$ [97], showing that SNase is one of the few proteins that are not completely unfolded in 8 M urea. SNaseD, the truncated form of SNase lacking the C-terminal 13 residues, is devoid of secondary structure under physiological conditions but is rather compact $(R_g = 2$ nm); for most of its mutated forms, R_g was found to be around 1.7 nm, and for a few ones it is up to 3.5

nm [103]. Since $(R_g)_\theta = 4.1$ nm, mutated forms of SNaseD are probably all collapsed, but to different extents. Circular dichroism spectra suggest that the forms of SNaseD with the smallest R_g values contain larger amounts of secondary structure than less compact ones [103].

Earlier results suggesting that proteins retain some structure upon heat denaturation are reviewed in [2,3]. Certainly, some of them are not in fact conclusive, as picked up by Privalov [35]. In particular, secondary structures that have been observed in heat denatured proteins could result from the aggregation of the unfolded molecules: we shall see later that formation of intermolecular β-sheets is frequently implied in this process. Nevertheless, one cannot dismiss such results outright, all the more so as they receive some backup from recent work using other methods. It is demonstrated for example by NMR that the conformation of thermally denatured lysozyme is substantially nonrandom, probably because of some degree of hydrophobic collapse, and that the departure from the random conformation decreases in a similar way upon reduction of disulfide bonds in the denatured protein or by action of urea without removal of the cross-links [104]. These results are consistent with earlier ones based principally on viscosity, optical rotation, and difference spectrophotometry measurements [105,106]. Ribonuclease does seem to retain secondary structure upon heat denaturation under both nonreducing and reducing conditions, as shown recently by Fourier transform infrared spectroscopy [107]. Investigation of a series of mutants of SNase suggests that the differences observed in the characteristics of their thermal denaturation are to be ascribed to differences in their heat denatured states [108]. One or two amino acid substitutions only appear thus to induce large changes in the hydration of the denatured state [108]. The study on the size of the mutants of SNaseD already referred to [103] resulted in a similar conclusion.

At this point in the discussion, it is clear that unfolded proteins do not display one single type of conformation, the random coil or the fully unfolded one, but a large range of conformations differing in their compactness and/or in their degree of organization. Some of them appear considerably more expanded than the native state, but less than expected for a random coil, and are devoid of secondary structure; they can be viewed as random coils that have more or less collapsed. The others show some amount of secondary structure and are generally rather compact (although less than the native state). The so-called residual secondary structure could be induced by the collapse of the coil; this indeed would be in line with some theories of protein folding based on simulation results, indicating that increasing compactness of a chain molecule leads to formation of secondary structure [76]. However, there is evidence that in many cases denatured conformations showing secondary structure correspond in fact to intermediary states (in the thermodynamic sense), which become the predominant species under some solvent conditions, or after chemical or genetic modifications of the protein as hinted at in the previous sections. Such conformations would therefore differ from the ultimate denaturation states (random coil, or collapsed conformations possibly with secondary structure) in that the secondary structure they contain should represent a part of the native folding, and that they should be separated from the native and ultimate states by configurational transitions (energy barriers).

Freire et al. [63] have shown that for single-domain proteins, the thermodynamically most probable type of partially folded intermediates is the one that retains most elements of the native secondary structure and a compact conformation, but in which the native rigid and specific tertiary structure is disrupted. These are the distinctive features of the so-called molten globule state, first suggested as a folding intermediary by Ptitsyn, and later observed as a partially stable state of several proteins when submitted to acid denaturation, or to temperature-induced (heat or cold) denaturation at acid pH or in presence of intermediary concentrations of urea or GuCl [109–110]. The different aspects of the molten globule question have been extensively reviewed [109–114]; in recent years there has been special interest in the energetics aspects [47,115–118]. Two important points deserve special attention and will be developed briefly below.

Does the considerable secondary structure retained by the molten globule represent part of the native structure, as it should if the molten globules were actually identical to the thermodynamically stable kinetic intermediates observed upon protein folding before organization of the specific tertiary structure [107,111,112,119,120]? Hydrogen exchange and NMR studies of the structure of the supposedly equilibrium molten globule states of cytochrome c [121], apomyoglobin [122], and α-lactalbumin [123–125] point to a positive answer; however, the possibility that some nonnative structural elements are also present cannot be ruled out [125]. The results indicate besides that these intermediary compact states retain also part of the native fold and native hydrophobic core, although the long-range tertiary structure is likely to be nonspecific. This definitely gives body to the abstract concept of a molten globule.

Calorimetric studies on α-lactalbumin [115] and apomyoglobin [116] have confirmed that the transition from the native state to the molten globule one is a highly cooperative two-state process [109] and can be considered as a first-order intramolecular phase transition [112]. They showed, contrariwise, that the transition from the molten globule to the ultimate unfolded state is rather noncooperative and would correspond to a second-order (or higher-order) intramolecular phase transition. These results have been analyzed [115] in the light of the hierarchical cooperative model of Freire and Murphy [33] already referred to. The model accounts for the existence of multiple partially unfolded states of denatured proteins separated by very small enthalpy differences, showing different degrees of secondary structure and of compactness, among which the typical molten globule state would represent the one with the highest degree of residual structure and compactness, and which is the first reached. The study of GuCl denaturation of the heat shock protein DnaK by a variety of methods leads to the same conclusion [126]. The scheme

$$\text{Native} \xrightarrow{\text{1st-order}} \text{Molten globule} \xrightarrow{\text{2d-order}} \text{Coil:}$$

is indeed predicted for solvent denaturation of proteins by the theory of Shakhnovich and Finkelstein and is in agreement with Monte Carlo simulations of protein folding [127]. However, for other proteins the transition from a compact partially ordered denatured conformation to the completely denatured state, studied mainly by noncalorimetric methods, seems to be a two-state, first-order one [102], as for instance the acid denaturation of cytochrome c [128,129], the heat denaturation of native and mutated SNase and its large fragment 1-136 [130–132], or for KCl-induced partial refolding of alkali denatured BLM [102]. A first-order transition from the molten globule to the completely denatured state has been considered recently by Ptit-syn to be a rule for almost all single-domain proteins [112,133,134], contrary to the above generally accepted scheme [131,135], and in opposition to his own earlier, more balanced, views [109]. In fact there is some evidence suggesting that "molten globules" from differ-ent proteins do not represent a unique conformation but a family of related states with different degrees of residual structure and com-pactness [117,118]. Besides, all intermediary states with secondary structure do not necessarily belong to the molten globule type; other types of partially folded intermediates have been observed [112]. In particular, it has been shown recently that the 1-136 fragment of SNase cannot be considered as a molten globule in its initial state but is constituted by one of the two cooperative domains that compose whole SNase [131].

The conclusion of this short survey is that the hypothesis of com-plete unfolding, i.e., complete exposure of protein groups to the sol-vent in the denatured state, underlying the classical thermodynamical approach to protein denaturation, turns out to be verified only excep-tionally or as a limiting case. The degree of exposure depends on the protein, the denaturation process, and the solvent conditions. More-over, some proteins at least present different thermodynamically dis-tinct denatured states separated by configurational transitions. Con-trary to the classical view, multistep denaturation of small proteins is far from being an exceptional case, at least when in the initial condi-tions the stability of the native state is decreased, although transition from partially structured intermediates to the final denatured state can be difficult to discern in calorimetric experiments.

The multiplicity and the partially unfolded character of the dena-turational states of proteins, especially upon temperature-induced de-naturation, hamper any thermodynamical analysis of the data and their interpretation. In particular, evaluation of the conformational entropy terms is actually infeasible.

V. DENATURATION-RELATED ASPECTS OF HEAT-INDUCED AGGREGATION

Aggregation of globular proteins during heat denaturation remains an ill-understood phenomenon. Its physical chemistry and its relation with gelation are beyond our scope; literature on the matter is limited [see Refs. 28, 136–139]. We shall restrict ourselves here to a few aspects directly linked to protein denaturation, which are in any case much better documented.

Formation of intermolecular disulfide bonds is frequently considered to be involved in the aggregation process and responsible for its irreversibility. Critical analysis of the abundant literature shows however that it plays only a marginal role.

BSA complies with the classical paradigm that protein unfolding is a reversible process. Oxidative "regeneration" of BSA at pH 8 after it has been unfolded in 10 M urea with reduction of its disulfide bonds restores the native conformation almost unaltered; the very slight differences in some properties observed with the native protein are due to the formation of polymeric material; the isolated refolded monomer does not differ actually from the native one [140]. A study on the temperature behavior of HSA at neutral pH led to a similar conclusion [141]. For solutions heated up to about 65°C the circular dichroism melting curves showed complete reversibility upon cooling. When solutions were heated at a temperature above 65°C, reversibility decreased the more the higher the temperature, the longer the time the solutions were kept at this temperature, and the higher the concentration. Since blocking the free SH groups by N-ethylmaleimide was found to prevent both irreversible aggregation and irreversible conformational changes, the authors suggested that irreversible aggregation proceeds through intermolecular disulfide bridges formation. However, the same study showed that this aggregation was paralleled with the development of β-sheet structures; moreover, aggregation was also observed to occur in dilute HSA solutions heated below 65°C, i.e., in conditions of complete reversibility of unfolding, with a concomitant substantial increase of β-structures in this case again; finally, it was observed that oligomers formed in the conditions of "irreversible" aggregation dissociated partially upon dilution after cooling (this was interpreted as resulting from the desorption of monomers bound to the aggregates). Therefore intermolecular disulfide bonds could well not be necessarily responsible for the formation of aggregates above 65°C; unfolded protein molecules might associate through the formation of intermolecular β-structures, and the apparently irreversible character of the phenomenon could be linked to the extent of aggregation. This extent and the strength of intermolecular interactions could be affected by changes in the status of sulfhyl and disulfide groups at the *intramolecular* scale. There is no reason why the aggregation mechanism above 65°C should differ in a clear-cut fashion from the aggregation mechanism below this rather moderate temperature.

Experimental evidence that SH blocking agents prevent the formation of aggregates in heated protein solutions have been frequently used to support the idea that intermolecular S-S bonds bridging via -SH/S-S interchange or -SH oxidation is responsible for heat-induced aggregation and gelation of globular proteins containing SH and S-S groups, such as, besides BSA or HSA, BLG and ovalbumin. However, this evidence, generally based on gel electrophoresis or size exclusion chromatography [23,142,143], could be interpreted as well in terms of an indirect effect along the same line we followed in the HSA case: blocking the SH groups could cause slackening of the aggregation process and therefore decrease the extent of aggregation for a given time of heating and make the aggregates more labile so that they could be dissociated when submitted to separation. One way in which blocking the free thiol groups of the proteins can affect their aggregation could be merely changing the electric charge of the molecules. The addition or suppression of one single charge by mutation was found to affect very significantly the temperature-induced polymerization of the protein of the tobacco mosaic virus [144,145]. On the other hand, modification of the SH groups of the native protein, which does not generally affect its conformation, could result in a change of the conformation of its denatured state, as we have seen to occur upon mutational changes of some proteins.

In the specific case of BLG, which has been probably the most studied protein in this respect, SH-blocking agents seem to act through other mechanisms.

BLG with its SH group blocked, although keeping its conformation and its dissociation behavior unchanged, was substantially less stable as regards denaturation by urea at neutral pH, the importance of the effect depending on the nature of the blocking agent [146]. Gotham et al. [147] suggest that blocking the SH group of BLG pre-

vents the formation, involving intramolecular S-S bonds, of a partially unfolded intermediary during heat denaturation at pH 6.6, in
accordance with the mechanism proposed by de Wit and Klarenbeek
[148].

In the neutral pH values range, BLG is mainly present as dimers
in solutions at room temperature. The sequence of events occurring
upon heating the solutions is not simple (see for example Refs. 149–
150); it is considered that the first step of the "reaction scheme" is
the dissociation of the dimer into native or quasi-native monomers,
followed by the denaturation of the two monomers; aggregation proceeds by association of dimers formed by denatured monomers [149].
Sawyer suggested that dimerization of the denatured monomers involves intermolecular disulfide bridging and is favored at high temperature [142]. Although these views cannot be considered as proven,
they seem indeed appealing, since they reconcile the possible role of
intermolecular disulfide bonds in the aggregation of BLG upon heat
denaturation with the probable noncovalent character of this aggregation on the whole.

What the work of Watanabe and Klostermeyer on BLG variant A
[143], often cited in support of the intermolecular S-S bond formation
mechanism in heat-induced aggregation of proteins, demonstrates actually is that a large part of the loss of sulfhydryl groups upon heating
the solutions at neutral pH in air is not reflected in a corresponding
increase in disulfide groups but must be due to the formation of small
sulphur compounds at the expense of -SH and S-S groups. It would
not be surprising if the chemical modifications of the protein that
such reactions imply had consequences for the aggregation properties
and the conformation of the denatured proteins. The fraction of -SH
plus S-S groups converted to other compounds increases sharply with
the temperature and the duration of the heat treatment [143]. This
could account for the current observations (cf. above) that heat denaturation shows often at least partial reversibility up to some temperature (about 70°C) but becomes irreversible as the solution is heated
above it. Besides, thiols being reactive in their ionized form only,
the effect of heating at 75°C on -SH and S-S levels becomes negligible below pH 6 [143]. As globular proteins in general and BLG in

particular do aggregate in solution and give gels upon heating at very acid pH values, intermolecular S-S formation cannot be a major factor in protein aggregation and its irreversible character. That SH-blocking agents do not prevent gelation is a supplementary argument. At this point one can conclude that aggregates of BLG are held by noncovalent interactions, although intermolecular S-S bridging could take place at the local scale, in forming doublets or small oligomers of denatured protein as a first step of the aggregation process for example. This limited participation of intermolecular disulfide bonds seems in addition to be rather specific to BLG (perhaps related to its dimeric character in native conditions at neutral pH), since it has not been observed in the case of ovalbumin for example. In their studies of heat-induced aggregation of ovalbumin at neutral pH by light scattering and size exclusion chromatography, Kato et al. found no effect of the presence of reducing or -SH blocking agents during heating on the molecular weight of the aggregates formed and showed that, once formed, the aggregates are readily dissociated by sodium dodecyl sulfate [151–152].

The noncovalent character of heat-induced aggregation of globular proteins is consistent with the fact that heat set gels of BLG [28] or of BSA [97] behave rheologically as viscoelastic liquids, meaning that the large scale connectivity of the network is achieved by noncovalent cross-links. This does not exclude the possibility that intermolecular S-S bonds could take place and contribute but at a local scale only. Anyway, extensive cross-linking of gels by disulfide bonds is unlikely, because the relatively compact conformation of heat-unfolded proteins leaves probably most of the cysteine residues unable to contract intermolecular bonds; formation of a covalent network would require an average of more than two "intermolecularly reactive" cysteine residues per molecule. The situation could be different when BSA or ovalbumin gelation occurs at high concentrations of urea or GuCl at alkaline pH, conditions insuring high -SH accessibility and chemical reactivity and high molecular flexibility; see for instance [153–155].

Heat-induced aggregation of globular proteins seems often to involve the formation of intermolecular β-sheets. We have mentioned above this fact in the case of HSA. It has been observed by infrared and laser Raman spectroscopy for a number of proteins [156]. It is well documented for instance in the case of ovalbumin [157–158] and S-ovalbumin [159]; results from Raman spectroscopy [158], from Fourier transform infrared spectroscopy [159], and from circular dichroism [157] converge.

Two questions bearing a direct relation to the subject of this chapter are raised by the preceding discussions:

—Could the secondary structure frequently observed in heat denatured proteins result in some cases from aggregation and not be a residual one? One could imagine a mechanism analogous to the "hydrophobic zipper" hypothesis of Dill et al. [160] acting at the intermolecular level.

—Does heat-induced aggregation involve in all cases "unfolded" protein molecules? In Sec. II we saw that BSA and ovalbumin have a definite tendency to aggregate at room temperature; this could be also the case for HSA [141] and RNase [161], since the heat stability of these proteins depends on concentration. Does this tendency increase with temperature below the temperature of incipient unfolding? How could this aggregation process involving native species affect heat-induced unfolding?

These are interesting basic problems waiting for specific investigation, but certainly difficult to tackle experimentally. Let us go back to more practical considerations.

VI. THERMAL DENATURATION OF PROTEINS: SOME EXPERIMENTAL PROBLEMS

Because of its theoretical and practical importance, the thermal behavior of proteins in aqueous media has been the subject of many investigations; but the results reported in the literature are often divergent, nay inconsistent, to say nothing of the interpretations of the data. The reason is that in practice many difficulties can be encountered in such studies that are frequently overlooked.

Section IV pointed to the problems related to the existence of different possible denatured states that are poorly characterized in general. But the initial state of the protein may be also ill-defined. We have seen (Sec. II) that in the native state proteins can exist under several forms that can show different behaviors with respect to heat denaturation.

When dealing with samples obtained by large-scale preparative procedures, and especially with industrial preparations, the protein can be contaminated with different substances that might affect its denaturation properties. In addition, it can have been submitted to different physical and chemical treatments, with the consequence that all or part of the protein in the sample is no longer in its native state; because of spatial gradients and time fluctuations of the preparation conditions, and since these conditions can cause irreversible changes, a given sample is not homogeneous as regards the denaturation state, and reproducibility between batches can be poor. Composition differences apart, the main cause of variability between samples is aggregation, which does not require drastic conditions to occur contrarily to chemical modifications. Aggregation increases with concentration. Rather high protein concentrations are precisely those of practical interest from the technologist's point of view. For the same reason, more or less (often rather more than less) extensive aggregation of the proteins during the scans is also a major factor responsible for the discrepancies in the results obtained with classical DSC instruments, which work in the 10–100 mg/mL protein concentration range. In such conditions, beside the fact that Q_D and T_m are then apparent quantities and depend on concentration and scan rate (hence hampering the comparisons of literature data), the effect of any factor on DSC data, for example an apparent stabilizing or destabilizing effect of a salt, could result from an action either on protein stability or on the aggregation process, or on both. However in spite of these limitations it remains possible to deduce from the characteristics of the heat capacity peak useful parameters that can be interpreted in terms of "thermal stability" and can be used for quality control in food technology.

Let us illustrate some of these points in more detail with a few experimental results selected from an abundant literature, without aiming at an exhaustive inventory, which would prove a formidable task.

A. Effects of Microheterogeneity in the Native State

1. Effect of Microheterogeneity Due to Ligand Binding: The BSA Case

In Sec. II we mentioned that the lipids bound by BSA in the physiological conditions affect its heat denaturation behavior. In certain conditions at least, the protective effect of fatty acids can prove indeed remarkable; according to the authors, the binding of fatty acids prevents the formation, by intermolecular SH/SS exchange reaction upon heating, of a BSA modified monomer, which would be a first step necessary for aggregation to take place [162]. The modified BSA monomer is partially unfolded, with contents in α-helix and β-structure slightly lower and higher, respectively, as compared to the native protein [163]. Using BSA complexes with radioactive labelled palmitic acid, the same research group showed that the fatty acid is released upon unfolding of BSA and therefore must concentrate on a particular species of BSA molecules that resists unfolding [164]. One should use defatted serum albumins when studying the heat denaturation of these proteins; this does not seem to be always the case in the literature.

To our knowledge, only one paper has been published on the calorimetric investigation of the effect of temperature on BSA–fatty acid complexes [165], showing that in solution T_m and Q_D increased with fatty acid content. On the other hand, an impressive stabilizing effect of SDS binding on defatted BSA was observed by DSC in the range of the SDS/protein molar ratio where SDS binding does not alter significantly the native conformation of the protein; it was explained by a bridging effect of the lauryl anion, its hydrophobic tail and anionic head binding to a nonpolar residue of the protein surface and to an adjacent positively charged group, respectively [166]. However one cannot rule out that the effect of SDS binding on BSA aggregation could have contributed to some extent to these results; in absence of SDS, preheating was observed to decrease the DSC peaks [166], indicating that the denaturation is only partially reversible and therefore probably accompanied by aggregation; since binding of SDS increases the net charge of BSA, it should decrease the propensity to aggregation. Besides fatty chains, a great variety of ligands have a protective effect on BSA and HSA against heat denaturation [167–168].

2. Effects of Other Types of Microheterogeneity

It has been observed that purified serum albumin from aged mice is more susceptible to heat denaturation than that of adult mice, the reason being unknown [169]. Another example of a biological aging phenomenon affecting protein stability is the development of the S-form of ovalbumin during storage of eggs. This form is characterized by its heat stability, demonstrated by DSC, T_m being as high as 8°C above that of ovalbumin at pH 9 [13]. The conversion of ovalbumin to S-ovalbumin, which is irreversible and proceeds through an intermediary [13], remains an enigmatic process. It involves neither sulfhydryl and phosphoseryl groups nor the loss of amino acids or of a peptide. The two forms differ slightly by their surface charges [170,171], by their surface hydrophobicity, which is higher for the S-form, and by their overall size or shape [172], but very little, if any, by their secondary structure [158,170,173]. Ovalbumin aggregates more rapidly than the S-form [174–175]. Accordingly, gels of S-ovalbumin show lower viscoelasticity than gels of ovalbumin obtained in the same conditions [176]. Samples of ovalbumin will contain the two forms in variable proportions, and the proportions within the same sample will change upon storage, since it was observed that partial conversion occurred even in freeze-dried preparations [13], making comparisons difficult.

Denaturation of ovalbumin by urea or GuCl was long considered to show several kinetic steps and poor reversibility [177–178]. Microheterogeneity had been suspected to be the cause [177]; this was confirmed later, when GuCl denaturation of the purified variant of ovalbumin containing two phosphate groups (the major constituent) was shown to conform to the classical two-state reversible transition process [179].

At pH 2, BLG is monomeric; its heat-induced unfolding is a two-state process, whereas its cold denaturation appears more complex [57]. At neutral pH BLG dimerizes and the monomer-dimer equilibrium is shifted to the left as temperature increases. It has been reported recently [180] that the DSC thermograms obtained on 0.4% BLG solutions at pH 7 are complex and show shoulders on the low-temperature side of the peak. The authors suggest that the main shoulder, situated in the 50–60°C region, reflects the dissociation of the dimer; unfolding will then occur in the monomeric form of the protein as considered in the usual models of BLG heat denaturation [149–150]. At high concentrations on the contrary, BLG would denature in the dimeric form; indeed, DSC scans performed at high protein concentrations show a nearly symmetrical and relatively narrow peak in the 70–80°C region.

B. Aggregation and Reversibility in DSC Studies

1. Effect of Aggregation on Thermograms

In most cases, in the course of DSC experiments the heat capacity peak due to unfolding and the one resulting from aggregation are overlapping, and one single resultant peak is observed. Aggregation of unfolded protein molecules is (1) a spontaneous, hence exothermic process; (2) a slow and continuously developing one at the difference of unfolding; (3) irreversible upon cooling in usual conditions; (4) the higher the larger the concentration of the protein is. The consequence of point 1 is that extensive aggregation during a DSC run will decrease the apparent value of Q_D. According to points 2 and 4 this decrease will be the higher the longer the protein is exposed to conditions leading to aggregation, i.e., the slower the heating rate in the calorimeter, and the larger the concentration. The apparent value of T_m should be also decreased, but an increase could perhaps occur also, depending on the relative importance of the two peaks, their shape, and their location on the temperature scale.

These effects are well illustrated by the work of Sochava et al. [161] on BSA and ribonuclease and of Relkin et al. [181,182] on BLG. For example, in the case of pH 7 BSA solution at the very high concentration of 35%, Q_D increased steadily from about 620 kJ mol^{-1} for a heating rate of 3 K min^{-1} to about 940 kJ mol^{-1} for a heating rate of 50 K min^{-1}, a value close to that determined in dilute solution [161]. Figure 3 shows that, as expected, Q_D and T_m decrease as BLG concentration increases at pH 3.2 [181]. At the lowest concentration (3.5%), Q_D was found to be 316 ± 25 kJ mol^{-1} [181], a value very close to that (312 kJ mol^{-1}) obtained by Griko and Privalov [57] for BLG variant A very dilute solutions (0.08–0.3%) at pH 2 and 0.1 M KCl. However, T_m was about 10°C higher (Fig. 3) than the value of 78°C obtained by Griko and Privalov. The anomalous T_m value of Relkin and Launay could be due either to the composition or to the state of their protein sample, which was prepared by ultrafiltration; one could think of a stabilizing effect of the salt or lactose it contained. The reason is perhaps simply that at pH >3 BLG solutions contain significant amounts of octamer even at temperatures higher than the ambient. Recently, $T_m = 84.2$°C and $Q_D = 380$ kJ mol^{-1} were obtained for the same BLG sample at pH 3.5 in 0.1 M NaCl [45]. These values agree with those found in similar conditions on BLG samples from Sigma, using high-sensitivity [183] as well as

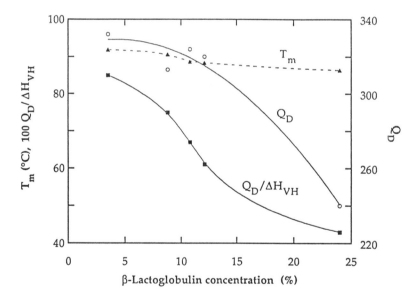

Figure 3 Effect of protein concentration on the heat denaturation of β-lactoglobulin at pH 3.2 monitored by DSC. Q_D: apparent calorimetric enthalpy of the transition; T_m: apparent transition temperature; ΔH_{VH}: Van't Hoff enthalpy of the transition. Heating rate: 10°C min^{-1}; heating from 20 to 120°C. (Data from Ref. 181.)

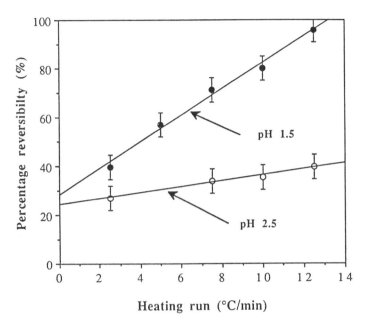

Figure 4 Effect of scan rate and pH on the percentage reversibility of peak transition observed with a β-lactoglobulin concentrate (4.25%, heating from 20 to 105°C). (Reported from Ref. 188.)

classical [148] DSC. They confirm the apparent high thermal stability of BLG between pH 3 and pH 5 already noticed by several authors [184–186], whatever the explanation could be.

2. Aggregation and Reversibility

Reversibility as measured by the ratio q of Q_D obtained in a second heating ramp, following the first one after rapid cooling, to Q_D obtained in the first run will be the higher in the conditions where aggregation is minimized (point 3). Thus for BLG at pH 2.3, $q = 0$ was found for a heating rate of 7.5°C min^{-1}, but q increased with the heating rate and reached 41% at 15°C min^{-1} [182]. As aggregation is opposed by electrostatic double layer repulsions, it will be minimized in solvent conditions which insure a high net charge of the protein and low screening of the electrostatic interaction [28,136,187]. No

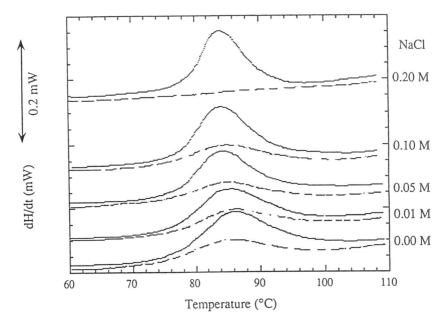

Figure 5 Effect of added NaCl on the transition peaks observed with a β-lactoglobulin isolate (4.9%, heating from 30 to 100°C at 10°C min^{-1}). Dashed curves: first heating; dotted curves: second heating, after annealing at 30°C for 4 min. (Reported from Ref. 189.)

wonder then if 75% reversibility is observed for BLG at pH 2 in 0.01 M NaCl, but only 56% when the salt concentration is 0.1 M; q drops similarly to 56% at pH 3 in 0.01 M NaCl and undergoes a further decrease at this pH value to 24% in 0.1 M NaCl [28]. Other results on BLG [188,189], illustrating the effect of heating rate, pH, and ionic strength on reversibility, are shown in Figs. 4 and 5; the samples in these cases were obtained by pilot plant-scale ion exchange chromatography and ultrafiltration, respectively.

In parallel to the decrease of the q ratio, aggregation causes a decrease in the ratio $r = Q_D/\Delta H(T)_{vH}$, as already mentioned. At low pH, BLG is monomeric and its heat-induced unfolding is a two-state process, with accordingly r very close to 1 as shown by high-sensitivity DSC in the 0.08–0.3% protein concentration range [57]. When experiments are performed at significantly higher concentrations in classical DSC instruments, values of r less than 1 are observed. As expected, they are the higher the lower the protein concentration [181] and the higher the heating rate [182]. Working at high heating rates appears attractive, but this is not free of risks, which are discussed in [161].

It is frequently observed in the study of heat denaturation, whatever the method used, but especially with DSC, that reversibility decreases with the increase of the final temperature reached during heating. There are two reasons for this:

—Increasing the final temperature means usually increasing the time necessary to reach it, and therefore the time during which the protein undergoes denaturational conditions with the consequence for aggregation that we have discussed;

—Higher temperatures favor chemical modifications (irreversible) of the protein; it is thus wiser to keep the final temperature as low as is compatible with the study of the denaturation transition, except of course if one were specifically interested in phenomena occurring at high temperature.

Chemical alteration of proteins under heating is particularly important at basic pH values; this has been shown for -SH and S-S groups by Watanabe and Klostermeyer in the paper already cited [143], but it is probably not limited to these groups. The net charge of BLG is about the same at pH 9 and at pH 2; however, in 0.01 M NaCl, q was observed to be zero at pH 9 against 75% at pH2 under the same conditions [28]; the difference is probably related to modification of the sulphur groups of the protein at alkaline pH.

An indisputable demonstration that thermal unfolding of globular proteins can be fairly reversible even at high protein concentrations, provided double-layer interactions between protein molecules are sufficiently strong, is given by the observation of an exothermal peak upon cooling the solution at the same rate after having heated it in a first temperature ramp. Complete reversibility has been found in this way for 15% RNase solution at pH 2.9 and low ionic strength under a heating and cooling rate as low as 1°C min^{-1} [161], and for α-lactalbumin in its apo and holo forms (Fig. 6) at pH 6.7 and 10% concentration [190]. In the case of a 5% BLG solution, 60% revers-

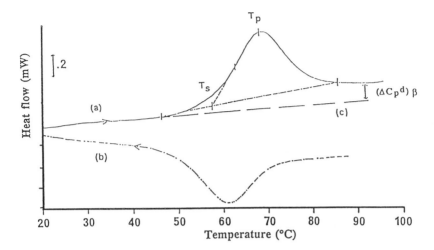

Figure 6 Heating (a) and cooling (b) thermograms obtained with α-lactalbumin (5% protein concentration, pH 6.5). ΔC_p^d is the heat capacity change between native and denatured forms; β is the heating scanning rate. (From Ref. 190, with permission.)

ibility was obtained at pH 2.3 with a temperature ramp of 5°C min^{-1} [182].

C. Specific Effects of Electrolyte Ions on Denaturation

Up to 0.1 M ionic strength, a general nonspecific stabilizing effect of salts, due to the screening of intramolecular electrostatic interactions, can be expected, but the same shielding action will promote aggregation, and the two phenomena are not separable. One observes actually that most ions tend to increase the apparent values of Q_D and T_m. Nevertheless the magnitude of the apparent stabilizing effect, for a given ionic strength, generally varies more or less with the nature of the ion (see [67], an example among a vast literature). The reason is that besides their screening effect at low ionic strength, the electrolyte ions interact weakly with accessible charged groups or polar residues of the protein, both in its native and its unfolded states, and this opens ways for selective effects on conformation, stability, and aggregation of the protein through selective binding [191]. Seemingly specific effects of ions on stability are much more pronounced at high salt concentrations [67,192], where normally stabilization of protein conformation is observed with cosmotropic anions and its destabilization with chaotropic ones [192].

The differences observed by DSC in the effects of the different ions at high ionic strength as well as at low ionic strength could be due to their effect on aggregation. It is generally found that they follow the famous Hofmeister series; the Hofmeister series was actually originally derived from the study of salting in/salting out effects [193], i.e., in the propensity of the different ions to promote dissociation or aggregation of proteins. Salt effects at high ionic strength are thought commonly to occur through modifications of the structure of water; chaotropic ions would tend to disorganize the structure of water and cosmotropic ones would have the opposite action; however, the ranking of the ions is frequently found to be the same as at low electrolyte concentrations [191]. Qualitatively similar effects of the ions are observed on the solubility of small organic dipolar ions and even nonpolar nonelectrolytes (see for instance [194]), so that in general the specificity of the ions lies actually rather in their intrinsic properties than in their effect on protein thermal behavior; in this respect, the word "specificity" is somewhat misleading. Keeping these reservations in mind, which apply to most DSC studies on salt effects on protein denaturation, the specific effects of some ions on heat stability of some proteins seem nevertheless to be ascertained. This is the case of course when the contribution of aggregation can be considered as negligible because of the low protein concentration; for example, Griko and Privalov [57] show a stabilizing effect of phosphate ion on BLG A variant (concentration ~0.1%) at pH 2. Even at high protein concentration, when aggregation interferes, evidence for specific effects of a given ion on the thermal stability of a particular protein can stem from a departure from what is observed usually on other proteins. Damodaran [192] found that $NaClO_4$ destabilizes BLG, as expected from the chaotropic character of ClO_4^-, but on the contrary has a stabilizing effect on BSA. Confirmation of a specific effect of ClO_4^- on BSA came from the observation that this anion increases the helix content of BSA, whereas it does not change the far-UV dichroic spectrum of BLG [192]. Aoki et al. [67], using

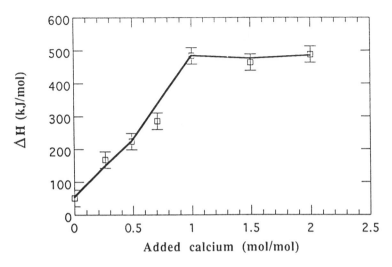

Figure 7   Variation of enthalpy change of heat denaturation of α-lactalbumin in apo form, as a function of Ca^{2+}/protein molar ratio. (Data from Ref. 190.)

DSC, reported that SCN^-, another chaotropic anion, stabilizes BSA in the same conditions. It is known that SCN^- binds strongly to BSA [191]. We observed [195] that Ca^{2+} and Mg^{2+} decrease Q_D and T_m of ribulose 1,5-diphosphate carboxylase (Rubisco) to a limited extent but in a definite and steady way as their concentration increases over the ionic strength range 0–0.3 mol/L; Na^+ and NH_4^+ had an opposite effect. In the case of Rubisco, specificity of the effect of Ca^{2+} and Mg^{2+} is supported by the fact that these cations are known to bind to the protein and to affect its enzymatic activity by inducing structural changes. Specific effects of Na^+ and Ca^{2+} are also observed by DSC on the heat stability of α-lactalbumin [190]; they result in a linear increase of Q_D with the cation/protein molar ratio up to 1, followed by a plateau (Fig. 7).

VII. INTERFACIAL DENATURATION AND ITS IMPORTANCE TO INTERFACIAL BEHAVIOR

Emulsions and foams are, intrinsically, thermodynamically unstable systems; they remain so in the presence of emulsifying or foaming agents. Adsorption of these agents does not suppress in general the interfacial tension, which measures the excess free energy of the system associated with the interfacial area, although it does decrease it; the stabilization it confers is of a kinetic nature: the adsorption layer formed slackens the rate of the destabilization processes by opposing to them electrostatic or (and) steric energy barriers. In the case of the classical DLVO stabilization mechanism, coagulation and therefore coalescence is hindered by interdroplet double-layer repulsions; the efficiency of stabilization is governed primarily by the net charge conferred on the "particles" by the adsorbed layer and by the ionic strength in the continuous phase. The thickness, the rheology, and the interactions with the phases of the polymer adsorption layer, i.e., its structure, are involved in the case of the polymer-type steric stabilization mechanism. Because of the macroion nature of proteins, both mechanisms are implied in their role as emulsifying and foaming agents.

The initial adsorption rate of protein molecules onto the interfaces is controlled by their transport properties; in unstirred conditions, it is diffusion limited, and diffusion limitation holds as long as there is a significant gradient of protein concentration near the interface. As proteins are macromolecules, their diffusion and consequently their adsorption are relatively slow processes. Still slower are the events taking place at the interface after the initial anchorage of the molecules, such as lateral mobility of protein molecules and conformational changes, which result in the formation of an "efficient" interfacial layer; they involve global or segmental mobility in conditions of spatial constraint and high concentration (supposing the interfacial layer is 100 Å thick, an interfacial concentration of 10^{-3} g/m^2 would be equivalent to a protein volumic concentration of 10^2 g/L in the layer). Emulsions and foams are usually formed under dynamic conditions; there will be a competition between the destabilizing processes at work and the development of a protein adsorption layer able to coat the droplets or the bubbles efficiently enough. Therefore kinetic aspects of adsorption are very important as to the practical performance of proteins as emulsifying and foaming agents; empirical indices used to evaluate globally protein efficiency, such as the so-called emulsifying capacity and emulsifying activity indices, in fact reflect partly the incidence of the adsorption rate [196]. However, they will not be considered here. We shall restrict ourselves to equilibrium aspects of adsorption; the aim is to stress a few points relative to the structure of protein adsorption layers, and not to give a review of the literature on the interfacial behavior of proteins; referring will be kept to a minimum and not systematic.

Adsorption of a protein onto a surface or interface can be split up into two processes, the initial anchoring of the molecule and a subsequent conformational change, although they are not actually observable separately. Understanding of both remains poor in depth. The first one involves probably hydrophobic patches existing on the surface of native globular proteins; in fact, a large proportion of the nonpolar side chains is not buried inside the protein, in spite of the tendency of nonpolar residues to be packed in the core of the native molecule [197]. Location of such patches at the contact of air or oil is energetically advantageous when considering the whole two-phase system, hence the "trapping" of the protein molecule at the interface; by the way, competition could be expected with aggregation in the aqueous phase, but aggregation of native proteins becomes perceptible only at much higher concentrations (cf. Sec. II) than the bulk concentrations usually involved in adsorption studies. It seems likely that an individual molecule anchored at the interface, in a nonisotropic environment thermodynamically completely different from the bulk, experiences an excess in free energy as compared to the same molecule in the bulk. However small, this excess would be sufficient to trigger unfolding of the protein molecule. In turn, the conformational change, which confers some degree of flexibility to the protein, would permit minimization of the free energy of both the system and the individual molecule by allowing the latter to distribute its groups in the most favorable way between the two phases and the interfacial plane, so that the molecule becomes strongly held at the interface. The energetic cost of desorption is then much higher than that of adsorption; this explains why desorption of proteins is in general difficult and their adsorption commonly seen as irreversible.

Albeit the whole spectrum of statements, ranging from no changes to complete unfolding, is found in the literature, that globular proteins undergo conformational changes upon adsorption is a generally accepted idea; but their extent remains a subject of discussion. Investigation of protein conformation at interfaces is indeed a much more difficult problem than in solution. In most cases, this is approached in an indirect way, from interfacial concentration and interfacial pressure isotherms; interpretation of such data, macroscopic by nature, in terms of molecular conformations is difficult and open to ambiguity. On the other hand, spectroscopic studies of proteins adsorbed at air–water or oil–water interfaces (infrared or circular dichroism are the most often used methods) are yet necessarily performed on films transferred onto solid supports by the Langmuir–Blodgett technique; actually, such films are likely to be very different from the interfacial layer. Spectroscopic studies on proteins adsorbed on solid surfaces do not meet this bias; but the nature of the interface affects certainly to a large extent the conformation of adsorbed proteins (this has been shown for solid surfaces with different hydrophilicities or hydrophobicities [198]), so that extrapolation from results obtained on solid surfaces to air–water and oil–water interfaces is not sound. However the following points find reasonable experimental support:

—Adsorbed globular proteins lose to a large extent their character of compact and rigid macromolecules (i.e., their tertiary structure), but they probably retain a corpuscular shape, in agreement with the soft particle model of de Feijter and Benjamins [199].

—They can keep secondary structure elements; this point is substantiated by numerous studies on synthetic polypeptides, which have demonstrated that α-helical or β-sheet polypeptides spread at air–water interfaces retain their conformations.

—Therefore, complete unfolding of the polypeptidic chain at the interface leading to a train-loop-tail model, similar to the conformation of copolymers at interfaces, is to be considered as a limiting case only. Even for a disordered and uncross-linked protein such as β-casein, this type of conformation at air–water or oil–water interface may not in fact be totally realistic [200]. It provides however a convenient basis to derive a thermodynamical analysis of protein interfacial layers from polymer theories [201–203]. Recent specular neutron reflectance studies on protein adsorption layers at the air–water interface [204] show that the protein density profile normal to the interface is qualitatively similar for β-casein and BLG and consists of a protein-rich layer, about 15 Å thick, close to the interface, and a diffuse layer, with a much lower volume fraction of protein, which extends far into the aqueous phase, down to ~80 Å from the interface plane for β-casein at pH 7 and to ~50 Å for BLG. Such profiles agree qualitatively with the train-loop model, the trains corresponding to the protein-rich layer, presumably constituted preferentially by the hydrophobic parts of the molecules, and the loops to the diffuse layer, corresponding to the more hydrophilic parts; but trains and loops are not necessarily disordered random segments as in the polymer model. The thicknesses and densities measured demonstrate without ambiguity that the globular protein adopts at the interface an expanded conformation, indicating a large degree of unfolding of the native structure, but this does not preclude that some elements of the secondary structure could be retained (if so, they would be probably located in the protein-rich layer and parallel to the interface plane, as expected from studies on α-helical or β-sheet synthetic polypeptides). Similar results were obtained previously by the same method for a spread monolayer of BSA [205]; the total thickness of the layer was significantly lower (~33 Å) than for BLG (~50 Å), meaning probably a lesser degree of unfolding of the former protein. Since the oil phase and the aqueous phase are rather good solvents for the hydrophobic and the hydrophilic parts of the protein molecule, respectively, adsorbed proteins probably unfold to a larger extent than heat-denatured proteins in solution.

—A wealth of results indicates that unfolding of adsorbed proteins depends not only on the nature of the interface but also on their native conformation and on the conditions in the aqueous phase.

The above points show that a parallel can be drawn between the unfolding of globular proteins in solution and that at interfaces. A recent paper suggests that it can be carried on even on quantitative grounds. Comparison of the wild type and six mutant α-subunits of tryptophan synthetase substituted at the same position has evidenced correlations between their surface properties at air–water interface, linked to their interfacial unfolding, and their free energy of unfolding in solution [206].

As in the case of protein denaturation in solution, interfacial unfolding can trigger aggregation of adsorbed protein molecules. When it occurs, adsorption becomes really irreversible, in the same way as aggregation caused irreversibility of protein denaturation in solution.

Because interfacial protein concentrations are very high, interfacial aggregation should be favored, but since molecular mobility is restricted at the interface it is likely to be a very slow process, to which perhaps could be ascribed the limited and slow evolution of interfacial pressure still observable many hours after stabilization of interfacial concentration when monitoring protein adsorption kinetics. In some conditions, interfacial aggregation can lead to interfacial coagulation [207,208] or gelation [209] of adsorbed proteins. But even when it does not go so far, aggregation modifies the rheological properties of the protein interfacial layer; it is probably responsible for the yield stress and the solidlike viscoelastic properties observed for example for BSA layers at dodecane–water interface [210]. These rheological properties of adsorption layers are directly and (or) indirectly implied in the stabilization of emulsions or foams. Intermolecular sulfhydryl–disulphide interchange can be involved in the aggregation of adsorbed globular proteins containing both SH groups and S-S bonds; this has been shown for BLG [211] and seems probable for ovalbumin.

We tentatively suggest that the frequently observed increase in stabilizing efficiency of proteins after they have been denatured in solution could be because they are aggregated and adsorb as aggregates, resulting in an interfacial layer that exhibits very quickly highly viscoelastic and solidlike rheological properties.

VIII. CONCLUSION

As stressed many times in this chapter, analysis of the denaturation of globular proteins is a complex problem. In practice, what is generally monitored is the global result of two different processes: the conformational transition of the molecule, and the subsequent aggregation of "unfolded" protein molecules; in many cases, the two processes cannot be separated, and this is a first difficulty in the thermodynamical approach. The second level of problems is inherent to the processes themselves considered separately.

Heat-induced conformational transitions of globular proteins have been the subject of an enormous amount of work from both the theoretical and the experimental points of view. Elaborated quantitative thermodynamical models have been built and "tuned to" structural data. Nevertheless, to achieve this goal a major difficulty was circumvented: it has been considered that temperature caused a complete unfolding of the protein molecule, i.e., a total exposure of all its groups to the solvent, although there was much experimental evidence that "complete unfolding" upon heat denaturation was indeed exceptional. The problem is now clearly recognized by theoreticians. But since the "unfolded" conformations are multiple and never precisely characterized, any thermodynamical analysis of protein unfolding is actually arbitrary, even in conditions where aggregation does not significantly interfere in the study.

Heat-induced aggregation of globular proteins remains a scarcely studied and ill-understood phenomenon in spite of its importance as regards the functional properties. Its investigation is by no means an easy task, since it requires us to work at supramolecular scales and in conditions of poor solubility. The main problem, however, is that the conformation of the species taking part in this process is actually unknown in general. It appears that careful and systematic application of DSC could make it possible, at least in favorable experimental conditions and for some proteins, to determine the thermal characteristics of both thermal unfolding and aggregation of unfolded molecules separately; this possibility of studying the aggregation process does not appear to have been really exploited yet.

The same kinds of problems are encountered with interfacial denaturation, the other type of denaturation of great technological importance. It poses however still greater difficulties to experimental and theoretical studies. Reliable characterization of the conformation of proteins adsorbed at air–water or oil–water interfaces would require the development of methods of working in situ; very useful information could be gained from systematic studies in neutron or x-ray reflectometry, but there is also a need for spectroscopic methods operating without film transfer. As to the interactions between adsorbed molecules and the interfacial aggregation phenomena, which play certainly a key role in emulsion and foam stabilization, insight can be expected from progress in interfacial rheology in shear.

REFERENCES

1. C. Ghelis and J. Yon, *Protein Folding*, Academic Press, New York, 1982.
2. C. Tanford, *Adv. Protein Chem. 23*: 121 (1968).
3. S. Lapanje, *Physicochemical Aspects of Protein Denaturation*, John Wiley, New York, 1978.
4. P. E. Stein, A. G. W. Leslie, J. T. Finch, and R. W. Carrell, *J. Mol. Biol. 221*: 941 (1991).
5. X. M. He and D. C. Carter, *Nature 358*: 209 (1992).
6. X. M. He and D. C. Carter, *Nature 364*: 362 (1993).
7a. T. Peters, Jr., *Adv. Protein Chem. 37*: 161 (1985).
7b. D. C. Carter and J. X. Ho, *Adv. Protein Chem. 45*: 153 (1994).
8. S. G. Hambling, A. S. McAlpine, and L. Sawyer, in *Advanced Dairy Chemistry- 1: Proteins* (P. F. Fox, ed.), Elsevier Applied Science, London, 1992, pp. 141–188.
9. G. Taborsky, *Adv. Protein Chem. 28*: 1 (1974).
10. F. Soetewey, M. Rosseneu-Motreff, R. Lamote, and H. Peeters, *J. Biochem. 71*: 705 (1972).
11. N. Sharon and H. Lis, in *The Proteins*, Vol. V (H. Neurath and R. L. Hill, eds.), Academic Press, New York, 1982, p. 7, p. 47.
12. D. T. Osuga and R. E. Feeney, in *Food Proteins* (J. R. Whitaker and S. R. Tannenbaum, eds.), AVI, Westport (Connecticut), 1977, p. 222.
13. J. Donovan and C. J. Mapes, *J. Sci. Food Agric. 27*: 197 (1976).
14. H. Pessen, J. M. Purcell, and H. M. Farrell, Jr., *Biochim. Biophys. Acta 828*: 1 (1985).
15. C. Georges and S. Guinand, *J. Chim. Phys. Physico-Chim. Biol. 57*: 606 (1960).
16. C. Georges, S. Guinand, and J. Tonnelat, *Biochim. Biophys. Acta 59*: 737 (1962).
17. M. Dupont, *C. R. Acad. Sci. 254*: 3137 (1962).
18. S. N. Timasheff, R. Townend, and L. Mescanti, *J. Biol. Chem. 241*: 1863 (1966).
19. R. Townend, T. F. Kumosinski, and S. N. Timasheff, *J. Biol. Chem. 242*: 4538 (1967).
20. S. N. Timasheff and R. Townend, *J. Am. Chem. Soc. 83*: 470 (1961).
21. P. Gough and R. Jenness, *J. Dairy Sci. 45*: 1033 (1962).
22. M. Dupont, *Biochim. Biophys. Acta 94*: 573 (1965).
23. M. McSwiney, H. Singh, O. Campanella, and L. K. Creamer, *J. Dairy Res. 61*: 221 (1994).

24. T. Matsumoto and J. Chiba, *J. Chem. Soc. Faraday Trans.* 86(16): 2877 (1990).
25. T. Matsumoto and H. Inoue, *J. Colloid Interf. Sci.* 160: 105 (1993).
26. T. Matsumoto and H. Inoue, *Chem. Phys.* 178(1–3): 591 (1993).
27. W. Burchard, in *Physical Techniques for the Study of Food Biopolymers* (S. B. Ross-Murphy, ed.), Blackie Academic and Professional, London, 1994, pp. 151–213.
28. D. Renard, Ph.D. thesis, University of Nantes (France), 1994.
29. N. Hagolle, unpublished results.
30. C. Tanford, *Adv. Protein Chem.* 24: 2 (1970).
31. C. N. Pace, *Crit. Rev. Biochem.* 3: 1 (1975).
32. T. E. Creighton, *Prog. Biophys. Molec. Biol.* 33: 231 (1978).
33. E. Freire and K. P. Murphy, *J. Mol. Biol.* 222: 687 (1991).
34. K. P. Murphy and E. Freire, *Adv. Protein Chem.* 43: 313 (1992).
35. P. L. Privalov, *Adv. Protein Chem.* 33: 167 (1979).
36. P. L. Privalov, *Adv. Protein Chem.* 35: 1 (1982).
37. J. F. Brandts, C. Q. Hu, L.-N. Lin, and M. T. Mas, *Biochem.* 28: 8588 (1989).
38. A. Ben-Naim, in *Protein-Solvent Interactions* (R. B. Gregory, ed.), Marcel Dekker, New York, 1995, pp. 387–420.
39. R. L. Biltonen and E. Freire, *CRC Crit. Rev. Biochem.* 5: 85 (1978).
40. P. L. Privalov and S. A. Potekhin, *Methods Enzymol.* 131(L): 4 (1986).
41. P. L. Privalov and S. J. Gill, *Adv. Protein Chem.* 39: 191 (1988).
42. P. L. Privalov, *Thermochim. Acta 163*: 33 (1990).
43. E. Freire, *Methods Enzymol.* 240(B): 502 (1994).
44. E. Freire, in *Protein Stability and Folding. Theory and Practice* (B. A. Shirley, ed.), Humana Press, Totowa (New Jersey), 1995, pp. 191–218.
45. P. Relkin, *Thermochim. Acta 246*: 371 (1994).
46. M. Straume, in *Methods Enzymol.* 240(B): 530 (1994).
47. D. Xie, V. Bhakuni, and E. Freire, *Biochem.* 30: 10673 (1991).
48. M. Straume and E. Freire, *Anal. Biochem.* 203: 259 (1992).
49. W. Pfeil, *Mol. Cell. Biochem.* 40: 3 (1981).
50. W. Pfeil, in *Biochemical Thermodynamics* (M. N. Jones, ed.), 2d ed., Elsevier, Amsterdam, 1988, pp. 53–99.
51. G. Castronuovo, *Thermochim. Acta 193*: 363 (1991).
52. Z. Dzakula and R. K. Andjus, *J. Theor. Biol. 153*: 41 (1991).
53. P. L. Wintrode, G. I. Makhatadze, and P. L. Privalov, *Proteins Struct. Funct. Genet. 18*: 246 (1994).

54. M. Renner, H.-J. Hinz, M. Scharf, and J. W. Engels, *J. Mol. Biol.* *223*: 769 (1992).
55. J. F. Brandts, *J. Am. Chem. Soc.* *86*: 4291 (1964).
56. P. L. Privalov, *Crit. Rev. Biochem. Mol. Biol.* *25*: 281 (1990).
57. Y. V. Griko and P. L. Privalov, *Biochem.* *31*: 8810 (1992).
58. Y. V. Griko, S. Y. Venyaminov, and P. L. Privalov, *FEBS Lett.* *224*: 276 (1989).
59. E. Freire, K. P. Murphy, J. M. Sanchez-Ruiz, M. L. Galisteo, and P. L. Privalov, *Biochem.* *31*: 250 (1992).
60. A. Horovitz, L. Serrano, B. Avron, M. Bycroft, and A. R. Fersht, *J. Mol. Biol.* *216*: 1031 (1990).
61. A. Horovitz and A. R. Fersht, *J. Mol. Biol.* *224*: 733 (1992).
62. C. N. Pace, G. R. Grimsley, J. A. Thomson, and B. J. Barnett, *J. Biol. Chem.* *263*: 11820 (1988).
63. K. P. Murphy, V. Bhakuni, D. Xie, and E. Freire, *J. Mol. Biol.* *227*: 293 (1992).
64. P. Relkin, A. Muller, and B. Launay, in *Food Macromolecules and Colloids* (E. Dickinson and D. Lorient, eds.), Royal Society of Chemistry, Cambridge, 1995, pp. 167–170.
65. P. L. Privalov and D. R. Monaselidze, *Biofizika 8*: 420 (1963).
66. M. Yamasaki, H. Yano, and K. Aoki, *Int. J. Biol. Macromol.* *12*: 263 (1990).
67. M. Yamasaki, H. Yano, and K. Aoki, *Int. J. Biol. Macromol.* *13*: 322 (1991).
68. P. L. Privalov and G. I. Makhatadze, *J. Mol. Biol.* *224*: 715 (1992).
69. G. I. Makhatadze and P. L. Privalov, *J. Mol. Biol.* *232*: 639 (1993).
70. P. L. Privalov and G. I. Makhatadze, *J. Mol. Biol.* *232*: 660 (1993).
71. G. I. Makhatadze and P. L. Privalov, *Biophys. Chem.* *51*: 291 (1994).
72. K. P. Murphy, *Biophys. Chem.* *51*: 311 (1994).
73. K. P. Murphy, P. L. Privalov, and S. J. Gill, *Science 247*: 559 (1990).
74. R. S. Spolar, J.-H. Ha, and M. T. Record, Jr., *Proc. Natl. Acad Sci. USA 86*: 8382 (1989).
75. G. D. Rose and R. Wolfenden, *Ann. Rev. Biophys. Biomol. Struct.* *22*: 381 (1993).
76. K. E. Dill, *Biochem.* *29*: 7133 (1990).
77. A. Hvidt and P. Westh, in *Protein Interactions* (H. Visser, ed.), VCH, Weinheim, 1992, pp. 327–343.
78. A. Nicholls, K. A. Sharp, and B. Honig, *Proteins Struct. Funct. Genet.* *11*: 281 (1991).

79. K. A. Sharp, A. Nicholls, R. F. Fine, and B. Honig, *Science 252*: 106 (1991).
80. A. A. Rashin, *Prog. Biophys. Molec. Biol. 60*: 73 (1993).
81. E. E. Lattman and G. D. Rose, *Proc. Natl. Acad. Sci. 90*: 439 (1993).
82. K. A. Dill, *Science 250*: 297 (1990).
83. P. L. Privalov, S. J. Gill, and K. P. Murphy, *Science 250*: 297 (1990).
84. K. H. Lee, D. Xie, E. Freire, and L. M. Amzel, *Proteins Struct. Funct. Genet. 20*: 68 (1994).
85. Y. V. Griko, G. I. Makhatadze, P. L. Privalov, and R. W. Hartley, *Protein Sci. 3*: 669 (1994).
86. Y. Yu, G. I. Makhatadze, C. N. Pace, and P. L. Privalov, *Biochem. 33*: 3312 (1994).
87. G. I. Makhatadze, G. M. Clore, A. M. Gronenborn, and P. L. Privalov, *Biochem. 33*: 9327 (1994).
88. J. M. Sturtevant, *Curr. Opin. Struct. Biol. 4*: 69 (1994).
89. A. R. Fersht and L. Serrano, *Curr. Opin. Struct. Biol. 3*: 75 (1993).
90. A. J. Doig and D. H. Williams, *J. Mol. Biol. 217*: 389 (1991).
91. C. N. Pace, D. V. Laurents, and J. A. Thomas, *Biochem. 29*: 2564 (1990).
92. J. Lefebvre, *Rheol. Acta 21*: 620 (1982).
93. W. G. Miller and C. V. Goebel, *Biochem. 7*: 3925 (1968).
94. C. Tanford, K. Kawahara, and S. Lapanje, *J. Am. Chem. Soc. 89*: 729 (1967).
95. S. Lapanje and C. Tanford, *J. Am. Chem. Soc. 89*: 5030 (1967).
96. M. Bohdanecky and V. Petrus, *Int. J. Biol. Macromol. 13*: 231 (1991).
97. J. Lefebvre, unpublished results.
98. D. F. Nicoli and G. B. Benedek, *Biopolymers 15*: 2421 (1976).
99. C.-C. Wang, K. Holland Cook, and R. Pecora, *Biophys. Chem. 11*: 439 (1980).
100. S. B. Dubin, G. Feher, and G. B. Benedek, *Biochem. 12*: 714 (1973).
101. G. Damaschun, H. Damaschun, K. Gast, D. Zirwer, and V. E. Bychkova, *Int. J. Biol. Macromol. 13*: 217 (1991).
102. Y. Goto and A. L. Fink, *Biochem. 28*: 945 (1989).
103. J. M. Flanagan, M. Kataoka, T. Fujisawa, and D. M. Engelman, *Biochem. 32*: 10359 (1993).
104. P. A. Evans, K. D. Topping, D. W. Woolfson, and C. M. Dobson, *Proteins Struct. Funct. Genet. 9*: 248 (1991).

105. M. Kugimiya and C. C. Bigelow, *Can. J. Biochem. 51*: 581 (1973).
106. K. C. Aune, A. Salahuddin, M. H. Zarlengo, and C. Tanford, *J. Biol. Chem. 242*: 4486 (1967).
107. T. R. Sosnick and J. Trewhella, *Biochem. 31*: 8329 (1992).
108. D. Shortle, A. K. Meeker, and E. Freire, *Biochem. 27*: 4761 (1988).
109. O. B. Ptitsyn, in *Protein Folding* (T. E. Creighton, ed.), W. H. Freeman, New York, 1992, pp. 243–300.
110. A. L. Fink, in *Protein Stability and Folding. Theory and Practice.* (B. A. Shirley, ed.), Humana Press, Totowa (New Jersey), 1995, pp. 343–360.
111. O. B. Ptitsyn, *J. Protein Chem. 6*: 273 (1987).
112. O. B. Ptitsyn, *Curr. Opin. Struct. Biol. 5*: 74 (1995).
113. K. Kuwajima, *Proteins Struct. Funct. Genet. 6*: 287 (1989).
114. K. A. Dill and D. Shortle, *Ann. Rev. Biochem. 60*: 795 (1984).
115. Y. V. Griko, E. Freire, and P. L. Privalov, *Biochem. 33*: 1889 (1994).
116. Y. V. Griko and P. L. Privalov, *J. Mol. Biol. 235*: 1318 (1994).
117. D. Xie and E. Freire, *Proteins Struct. Funct. Genet. 19*: 291 (1994).
118. D. Xie and E. Freire, *J. Mol. Biol. 242*: 62 (1994).
119. K. Kuwajima, M. Mitani, and S. Sugai, *J. Mol. Biol. 206*: 547 (1989).
120. P. A. Evans and S. E. Radford, *Curr. Opin. Struct. Biol. 4*: 100 (1994).
121. M.-F. Jeng, S. W. Englander, G. Elöve, A. J. Wand, and H. Roder, *Biochem. 29*: 10433 (1990).
122. F. M. Hughson, P. E. Wright, and R. L. Baldwin, *Science 249*: 1544 (1990).
123. J. Baum, C. M. Dobson, P. A. Evans, and C. Hanley, *Biochem. 28*: 7 (1989).
124. A. T. Alexandrescu, P. A. Evans, M. Pitkeathly, J. Baum, and C. M.Dobson, *Biochem. 32*: 1707 (1993).
125. C.-L. Chyan, C. Wormald, C. M. Dobson, P. A. Evans, and J. Baum, *Biochem. 32*: 5681 (1993).
126. D. R. Palleros, L. Shi, K. L. Reid, and A. L. Fink, *Biochem. 32*: 4314 (1993).
127. M. Karplus and E. Shakhnovich, in *Protein Folding* (T. E. Creighton, ed.), W. H. Freeman, New York, 1992, pp. 127–195.
128. M. Kataoka, Y. Hagihara, K. Mihara, and Y. Goto, *J. Mol. Biol. 229*: 591 (1993).

129. Y. Goto, Y. Hagihara, D. Hamada, M. Hoshino, and I. Nishii, *Biochem. 32*: 11878 (1993).
130. A. G. Gittis, W. E. Stites, and E. E. Lattman, *J. Mol. Biol. 232*: 718 (1993).
131. Y. V. Griko, A. Gittis, E. E. Lattman, and P. L. Privalov, *J. Mol. Biol. 243*: 93 (1994).
132. J. H. Carra and P. L. Privalov, *Biochem. 34*: 2034 (1995).
133. V. N. Uversky, G. V. Semisotnov, R. H. Pain, and O. B. Ptitsyn, *FEBS Lett. 314*: 89 (1992).
134. O. B. Ptitsyn and V. N. Uversky, *FEBS Lett. 341*: 15 (1994).
135. Z. Peng and P. S. Kim, *Biochem. 33*: 2136 (1994).
136. D. Renard, M. A. V. Axelos, and J. Lefebvre, in *Food Macromolecules and Colloids* (E. Dickinson and D. Lorient, eds.), Royal Society of Chemistry, Cambridge, 1995, pp. 390–399.
137. P. Kratochvil, P. Munk, and B. Sedlacek, *J. Polymer Sci. 53*: 295 (1961).
138. A. H. Clark and C. D. Lee-Tuffnell, in *Functional Properties of Food Macromolecules* (J. R. Mitchell and D. A. Ledward, eds.), Elsevier Applied Science, London, 1986, pp. 203–272.
139. A. H. Clark and S. B. Ross Murphy, *Adv. Polymer Sci. 83*: 57 (1987).
140. K. O. Johanson, D. B. Wetlaufer, R. G. Reed, and T. Peters, Jr., *J. Biol. Chem. 256*: 445 (1981).
141. R. Wetzel, M. Becker, J. Behlke, H. Billwitz, S. Böhm, B. Ebert, H. Hamann, J. Krumbiegel, and G. Lassmann, *Eur. J. Biochem. 104*: 469 (1980).
142. W. H. Sawyer, *J. Diary Sci. 51*: 323 (1968).
143. K. Watanabe and H. Klostermeyer, *J. Dairy Res. 43*: 411 (1976).
144. R. A. Shalaby and M. A. Lauffer, *Arch. Biochem. Biophys. 204*: 494 (1980).
145. R. A. Shalaby and M. A. Lauffer, *Arch. Biochem. Biophys. 204*: 503 (1980).
146. G. B. Ralston, *C. R. Trav. Lab. Carlsberg 38*: 499 (1972).
147. S. M. Gotham, P. J. Fryer, and A. M. Pritchard, *Int. J. Food Sci. Technol. 27*: 313 (1992).
148. J. N. de Wit and G. Klarenbeek, *J. Dairy Res. 48*: 293 (1981).
149. D. Pantaloni, *C. R. Acad. Sci. 259*: 1775 (1964).
150. W. G. Griffin and M. C. A. Griffin, *J. Chem. Soc. Faraday Trans. 89*: 395 (1993).
151. A. Kato, Y. Nagase, N. Matsudomi, and K. Kobayashi, *Agric. Biol. Chem. 47*: 1829 (1983).

152. A. Kato and T. Takagi, *J. Agric. Food Chem. 35*: 633 (1987).
153. C. Huggins, D. F. Tapley, and E. V. Jensen, *Nature 167*: 592 (1951).
154. H. K. Frensdorff, M. T. Watson, and W. Kauzmann, *J. Am. Chem. Soc. 75*: 5157 (1953).
155. F. S. M. Van Kleef, *Biopolymers 25*: 31 (1986).
156. A. H. Clark, D. H. P. Saunderson, and A. Sugget, *Int. J. Peptide Protein Res. 17*: 353 (1981).
157. A. Kato and T. Takagi, *J. Agric. Food Chem. 36*: 1156 (1988).
158. P. C. Painter and J. L. Koenig, *Biopolymers 15*: 2155 (1976).
159. T. J. Herald and D. M. Smith, *40*: 1737 (1992).
160. K. A. Dill, K. M. Fiebig, and H. Sun Chan, *Proc. Natl. Acad. Sci. USA 90*: 1942 (1993).
161. I. V. Sochava, T. V. Belopolskaya, and O. I. Smirnova, *Biophys. Chem. 22*: 323 (1985).
162. K. Aoki, N. Hayakawa, K. Noda, H. Terada, and K. Hiramatsu, *Colloid Polym. Sci. 261*: 359 (1983).
163. K. Aoki, S. Sakurai, M. Murata, T. Ito, H. Terada, and K. Hiramatsu, *Colloid Polym. Sci. 262*: 470 (1984).
164. K. Aoki, I. Nagai, K. Hiramatsu, *Int. J. Biol. Macromol. 6*: 293 (1984).
165. Y. Fu and Zh. Zhang, *Shengwu Huaxue Yu Shengwu Wuli Xuebao 16*: 140 (1984).
166. M. Yamasaki, H. Yano, and K. Aoki, *Int. J. Biol. Macromol. 14*: 305 (1992).
167. J. H. Brown and H. K. Mackey, *Proc. Soc. Exp. Biol. Med. 128*: 225 (1968).
168. K. Lohner, A. C. Sen, R. Prankerd, A. F. Esser, and J. H. Perrin, *J. Pharm. Biomed. Anal. 12*: 1501 (1994).
169. J. D. Schofield, *Exp. Gerontol. 15*: 533 (180).
170. R. Nakamura, M. Hirai, and Y. Takemori, *Agric. Biol. Chem. 44*: 149 (1980).
171. R. Nakamura, Y. Takemori, and S. Shitamori, *Agric. Biol. Chem. 45*: 1653 (1981).
172. R. Nakamura and M. Ishimaru, *Agric. Biol. Chem. 45*: 2775 (1981).
173. S. Kint and Y. Tominatsu, *Biopolymers 18*: 1073 (1979).
174. M. B. Smith and J. F. Back, *Aust. J. Biol. Sci. 21*: 539 (1968).
175. M. B. Smith and J. F. Back, *Aust. J. Biol. Sci. 21*: 549 (1968).
176. S. Shimatori, E. Kojima, and R. Nakamura, *Agric. Biol. Chem. 48*: 1539 (1984).

177. R. B. Simpson and W. Kauzmann, *J. Am. Chem. Soc. 75*: 5139 (1953).
178. J. C. Holt and J. M. Creeth, *Biochem. J. 129*: 665 (1972).
179. F. Ahmad and A. Salahuddin, *Biochem. 15*: 5168 (1976).
180. X. L. Qi, S. Brownlow, C. Holt, and P. Sellers, *Biochem. Soc. Trans. 23*: 74S (1995).
181. P. Relkin and B. Launay, *Food Hydrocolloids 4*: 19 (1990).
182. P. Relkin, L. Eynard, and B. Launay, *Thermochimica Acta 204*: 111 (1992).
183. F. P. Schwarz, *Thermochim. Acta 159*: 305 (1990).
184. N. K. D. Kella and J. E. Kinsella, *Biochem. J. 225*: 113 (1988).
185. P. O. Hegg, *Acta Agric. Scand. 34*: 401 (1980).
186. V. R. Harwalkar and C.-Y. Ma, in *Protein Interactions* (H. Visser, ed.), VCH, Weinheim, 1992, pp. 359–378.
187. D. Renard and J. Lefebvre, *Int. J. Biol. Macromol. 14*: 287 (1992).
188. T. Liu, P. Relkin, and B. Launay, *Thermochim. Acta 246*: 387 (1994).
189. T. X. Liu, Ph.D. thesis, Ecole Nationale Supérieure des Industries Alimentaires, Massy (France), 1993.
190. P. Relkin, B. Launay, and E. Eynard, *J. Dairy Sci. 67*: 36 (1993).
191. M. T. Record, Jr., C. F. Anderson, and T. M. Lohman, *Quart. Rev. Biophys. 11*: 103 (1978).
192. S. Damodaran, *Int. J. Biol. Macromol. 11*: 2 (1989).
193. F. Hofmeister, *Arch. Exp. Pathol. Pharmakol. 24*: 247 (1888).
194. J. T. Edsall and J. Wyman, *Biophysical Chemistry*, Vol. 1, Academic Press, New York, 1958, pp. 241–322.
195. V. Beghin, H. Bizot, M. Audebrand, J. Lefebvre, D. G. Libouga, and R. Douillard, *Int. J. Biol. Macromol. 15*: 195 (1993).
196. C. Dagorn-Scaviner, J. Gueguen, and J. Lefebvre, *J. Food Sci. 52*: 335 (1987).
197. C. Chothia, *Ann. Rev. Biochem. 53*: 537 (1984).
198. H. Elwing, S. Welin, A. Askendal, and I. Lundström, *J. Colloid Interface Sci. 123*: 306 (1988).
199. J. A. de Feijter and J. Benjamins, *J. Colloid Interface Sci. 90*: 289 (1982).
200. E. Dickinson, B. S. Murray, and G. Stainsby, in *Advances in Food Emulsions and Foams* (E. Dickinson and G. Stainsby, eds.), Elsevier Applied Science, London, 1988, pp. 123–162.
201. F. Uraizee and G. Narsimhan, *J. Colloid Interface Sci. 146*: 169 (1991).

202. R. Douillard, *Colloids and Surfaces B: Biointerfaces 1*: 333 (1993).
203. R. Douillard, M. Daoud, J. Lefebvre, C. Minier, G. Lecannu, and J. Coutret, *J. Colloid Interface Sci. 163*: 277 (1994).
204. P. J. Atkinson, E. Dickinson, D. S. Horne, and R. M. Richardson, *J. Chem. Soc. Faraday Trans. 91*: 2847 (1995).
205. A. Eaglesham, T. M. Herrington, and J. Penfold, *Colloids Surf. 65*: 9 (1992).
206. A. Kato and K. Yutani, *Protein Engineering 2*: 153 (1988).
207. F. MacRitchie and N. F. Owens, *J. Colloid Interface Sci. 29*: 66 (1969).
208. L. R. Fisher, E. E. Mitchell, and N. S. Parker, *J. Colloid Interface Sci. 119*: 592 (1987).
209. M. A. Cohen Stuart, J. T. F. Keurentjes, B. C. Bonekamp, and J. G. E. M. Fraaye, *Colloids and Surfaces 17*: 91 (1986).
210. C. Le Bihan and J. Lefebvre, in *Gums and Stabilisers for the Food Industry-6*. (G. O. Phillips, D. Wedlock, and P. A. Williams, eds.), IRL Press, Oxford, 1992, pp. 231–234.
211. E. Dickinson and Y. Matsumura, *Int. J. Biol. Macromol. 13*: 26 (1991).

8
Protein–Surfactant Interactions

Malcolm N. Jones

University of Manchester, Manchester, England

I. INTRODUCTION

The importance of the interaction between amphiphilic molecules and proteins dates back to the discovery by the Egyptians in the first century A.D. of the use of soap for personal washing. The soap molecules are brought into contact with the skin and hair surfaces and hence interact with the protein keratin. The scientific study of protein–amphiphile interactions appears to have its origins in the use of amphiphiles to detoxify snake venom and tetanus toxin by Fraser [1] and Phisalix [2] in 1897. By 1925 Larson et al. [3] had shown that 3% solution of sodium ricinoleate detoxicates diptheria toxin with the formation of an adsorption product. It was concluded that the toxin exists as a colloidal aggregate capable of being dispersed and surrounded by a protective soap layer. Following Larson et al., Baylis investigated the inactivation of diphtheria toxin by soaps and surfactants including sodium dodecylsulphate (SDS) in 1936 [4], and the field began to expand, notably with the study of the interaction of SDS with tobacco mosaic virus by Screenivasaya and Pirie in 1938 [5], and with studies of the interactions between commercial detergents and bile salts with methaemoglobin [6,7] and egg albumin [8] by Anson. The commercial detergents used by Anson were alkylsulphates manufactured by the DuPont Company under the trade names Duponol Special WA, which was mainly SDS and Duponol PC, a mixture of C_{10}–C_{18} sulphates. This early work firmly established that surfactants such as SDS could denature proteins at low concentrations relative to other known denaturants such as urea and guanidinium chloride. It should be noted that although the terms surfactant and detergent are often used interchangeably, rigorously a detergent is a formulation that contains among other components the surfactant as the active constituent.

The early work on the interactions of synthetic surfactants and fatty acid anions with proteins was reviewed by Putnam [9] in 1948 and the following 20 years by Steinhardt and Reynolds [10]. The expansion of the field has led more recently to reviews and monograph chapters dealing with specific areas of protein–surfactant interactions, such as the solubilization of membranes by detergents [11] and the use of surfactants in membrane solubilization and reconstitution of membrane proteins [12], as well as more general reviews [13,14,15].

From the viewpoint of surface activity, both components of a protein–surfactant system are individually surface active. The composition of proteins covers a wide range of hydrophobicities [16], and many proteins undergo denaturation when adsorbed at an interface (see Chap. 1). Interaction between proteins and denaturing ionic surfactants is initially electrostatic and occurs at the surface of the protein. Further interactions with denaturing surfactants exposes the hydrophobic amino acid residues, which in the native state of a protein are largely buried inside the globule. The resulting complexes are not appreciably surface active once all the binding sites have become saturated, although their solutions will have depressed surface tensions relative to water, as a consequence of the adsorption of surfactant molecules at the aqueous–air interface, with which the complexes are in equilibrium. The stoichiometry of saturated protein–surfactant complexes was first studied using equilibrium dialysis by Pitt-Rivers and Impiombato [17] and later by Reynolds and Tanford [18]. These workers established that for many globular protein–SDS complexes the proteins bound approximately 1.4 grams of SDS per gram of protein (i.e., approximately one SDS molecule per two amino acid residues) and that the interaction involved the "monomeric" surfactant as distinct from the micelles. Complex saturation by surfactants is usually complete as the "free" surfactant concentration in the solution approaches the critical micelle concentration (cmc). It follows that in general protein–surfactant complexes are more stable than the surfactant micelles and it is the stability at premicellar monomeric surfactant concentrations that leads to the widespread occurrence of the complexation of proteins by surfactants.

II. OCCURRENCE AND APPLICATION OF PROTEIN COMPLEXATION BY SURFACTANTS

There are two major areas in which surfactants find practical applications, in food technology and in the formulation of detergents and personal hygiene products. In both these areas the major functions of the surfactant are as emulsifiers and as solubilizing agents. Because proteins are major constituents of foodstuffs, the formation of protein–surfactant complexes is of major importance. In the processing and storage of food, surfactants perform a variety of functions including emulsification. Dickinson [19] identifies three main functions of emulsifiers as

1. Stabilization and partial destabilization of emulsions and foams by controlling the state of dispersion and agglomeration of oil droplets or fat globules
2. Modification of shelf-life, texture, and rheological properties through interaction with starch and protein components
3. Control of texture and morphology of fat-based products by influencing neutral lipid polymorphism, i.e., the nature and size of triglyceride crystals

In that most foodstuffs contain proteins, the formation of protein–surfactant complexes plays a direct role in 2. and an indirect role in 1. and 3., particularly with respect to their adsorption at oil–water interfaces and distribution between the phases. The surfactants used in foodstuffs are necessarily controlled, and permitted food emulsifiers are esters (or partial esters) formed from fatty acids (from animal and vegetable sources) with polyvalent alcohols such as glycerol, propylene glycerol, or sorbitol.

In the field of detergency and products for personal hygiene (soaps, shampoos, skin creams, toothpaste, etc.), surfactants are brought into contact with proteinacious soils and with the surface of the skin and hair. In the case of proteinacious soils on fabrics, for example, the formation of protein–surfactant complexes aids the solubilization of the soil, which is an essential part of the detergency process [20]. Detergency depends on the action of monomeric surfactant at the interface between the soil and the fabric or surface to be cleaned, and the formation of surfactant micelles does not aid this action. The fact that protein–surfactant complexes are formed prior to micellization, as the surfactant concentration is increased, is advantageous in removing and dispersing the soil.

Surfactants have found extensive applications in biochemistry, particularly in the area of membrane studies [11,12], where they are used to solubilize hydrophobic membrane proteins prior to their characterization and reconstitution in model membrane systems such as phospholipid liposomes (vesicles) or planar bilayer membranes. Many transport proteins, e.g., the anion and glucose transporters of the human erythrocyte, and many membrane receptors, e.g., the insulin receptor, the acetylcholine receptor, and the β-adrenergic receptor, have all been reconstituted in bilayer systems starting with nondenaturing surfactant solubilized functional proteins [12]. Surfactant binding to integral membrane proteins has been used as a measure of the hydrophobic surface of the proteins. From the volume of the bound surfactant and the surfactant dimensions, assuming the surfactant binds as a monolayer around the hydrophobic transmembrane domain of an integral membrane protein, the surface area of the hydrophobic domain can be estimated. The method has been applied to bacteriorhodopsin, the photosynthetic reaction center, sarcoplasmic reticulum Ca^{2+}-ATPase, and cytochrome oxidase [21].

In the field of research methods probably the most important application of protein surfactant complexation is in the technique of polyacrylamide gel electrophoresis in the presence of SDS, the so-called SDS-PAGE technique, used for the analysis and estimation of molecular masses of protein subunits [22]. Protein subunit–SDS complexes are formed from proteins reduced by β-mercaptoethanol to remove disulphide bonds. The binding of SDS to the polypeptide chains occurs uniformly along the chains so that the charge per unit length is constant, and the initial charge on the polypeptides is swamped by the bound surfactant. The complexes are rod-like, and because the charge per unit length is constant the electric force on them will be proportional to their length. In an electric field in the absence of a gel these complexes would all move at a uniform rate, because the frictional resistance to their motion, like the electric force on them, is proportional to their length. However in a medium of cross-linked polyacrylamide the complexes will be subjected to a sieving action, the larger complexes being slowed down relative to the small complexes. The result is that the gel affects a separation on the basis of molecular mass. Molecular masses can be estimated with reference to calibration standards of known mass run on the same gel.

III. CLASSIFICATION OF SURFACTANTS AND DENATURATION

Surfactants are generally classified on the basis of the character of their head groups, the principal categories being anionic (e.g., alkyl sulfates, alkyl benzene sulfonates, cholates), cationic (e.g., alkyl trimethyl ammonium halides, alkyl pyridinium halides), and nonionic (e.g., ethoxylated alkylphenols (Triton), alkyl glucosides, alkylpolyethylene glycols (Lubrols)). A more extensive classification would include zwitterionic (or amphoteric) materials such as the betaines (alkyl dimethyl ammonio methane carboxylate) and the zwittergents (alkyl-N, N-dimethyl-3-ammonio-1-propane sulfonates). More information on the types and properties of surfactants can be found in other volumes of the Surfactant Science Series (e.g., Ref. 23) and in the comprehensive list of surfactants and their critical micelle concentrations by Mukerjee and Mysels [24].

The most important division of surfactants with regard to their interaction with proteins is the distinction between surfactants that will unfold or denature proteins and those that will not. In that protein denaturation is a highly cooperative process, so that once unfolding has been initiated the denatured state inevitably follows, the division of surfactants into denaturing and nondenaturing is sharply defined.

We do not generally find surfactants that only partially denature proteins; although the course of denaturation involves the formation of a range of protein–surfactant complexes, the transition from native protein–surfactant complexes to denatured protein–surfactant complexes as a function of surfactant concentration is usually sharp (highly cooperative).

Ionic surfactants with alkyl chains are potent denaturants, i.e., they denature proteins at low concentrations. Thus sodium n-dodecylsulphate is a very potent protein denaturant; in general it will unfold most proteins at concentrations below its cmc ($\sim$8 mM in pure water). This contrasts with denaturants such as urea or guanidinium chloride, which denature proteins as a consequence of the changes they bring about in the structure of water and denature only at very high concentrations; 6–8 M for urea and 4–6 M for guanidinium chloride; SDS is thus a 500–1000 times more effective denaturant. There are however a few proteins that resist denaturation by SDS under some conditions; these include papain and pepsin [25], glucose oxidase near neutrality (pH 6.0) [26], and bacterial (*Micrococcus luteus*) catalase [27], while *Aspergillus niger* catalase near neutrality (pH 6.4) is activated by SDS but not in acid (pH 3.2) or alkaline solution (pH 10.0) [28]. Many synthetic ionic surfactants behave like SDS and can be classified as denaturing surfactants; the n-alkylsulphates, n-alkylsulphonates, n-alkyl trimethyl ammonium halides, and n-alkylpyridinum halides all act as denaturants. In contrast, the "natural" anionic surfactants, sodium cholates and deoxycholates, and nonionic surfactants do not denature proteins. Although the cholates and for example Triton X-100 and n-octylglucoside will bind to the tertiary structure of proteins, they will not unfold them, so that function is retained in the case of enzymes, membrane transporters, and receptors. There are also cases of activation by nondenaturing surfactants, e.g., Triton X-100 activates glucose-6-phosphatase [29], and sodium deoxycholate activates phospholipase [30].

The classification of surfactants into denaturing and nondenaturing poses the question as to the origin of the distinction between the two types of behavior. Since all surfactants have an amphipathic structure, why do synthetic ionic surfactants denature proteins while natural ionics and synthetic nonionics do not? The answer must clearly lie partially in the nature of the interactions between the head groups and the protein, at least in the initial stages of the interaction.

IV. THE CHARACTERIZATION AND MECHANISM OF SURFACTANT BINDING AND DENATURATION OF PROTEINS

The overall mechanism of surfactant-induced denaturation of globular proteins and the formation of protein–surfactant complexes is shown schematically in Fig. 1 as it relates to ionic denaturing surfactant. The initial step involves the binding of the surfactant head group to ionic sites on the surface of the protein. Anionic surfactants will bind to cationic sites (lysyl, histidyl, and arginyl residues), while cationic surfactants will bind to anionic sites (glutamyl and aspartyl residues). While head group interactions with charged sites on the protein surface are electrostatic, the surfactant alkyl chains interact hydrophobically with hydrophobic patches on the protein. The evidence for this type of initial interaction comes from binding isotherms. Figure 2 shows a schematic binding isotherm in which the average number of surfactant molecules per protein molecule ($\bar{v}$) is plotted as a function of the logarithm of the free surfactant concentration (log $[S]_{free}$). Region A corresponds to the specific binding of surfactant to charged sites on the surface of the native protein. When these sites

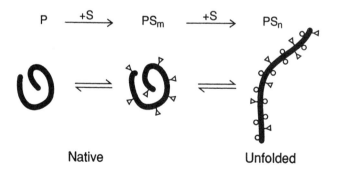

Native   Unfolded

Figure 1   The binding of surfactant ligands S to the native state of a protein P and surfactant induced unfolding. (Reproduced from Jones and Brass [97] with permission from the Royal Society of Chemistry.)

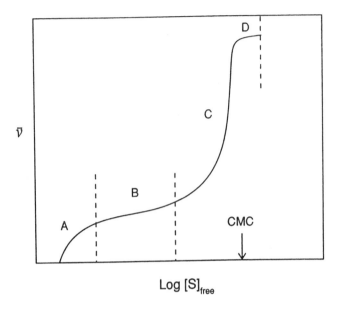

Figure 2 Schematic plot of the binding of surfactant ligands ($\bar{\nu}$ = number of ligands bound per protein molecule) as a function of the logarithm of the free surfactant concentration $[S]_{free}$. Region A corresponds to the specific binding region; B the plateau region; C the cooperative binding region, and D saturation in the region of the cmc of the surfactant.

are saturated, a plateau-like region occurs (Region B), which is followed by a steep rise as the surfactant binds cooperatively and hydrophobically to the unfolded protein (Region C). Region C generally occurs as the free surfactant approaches the critical micelle concentration. This region terminates when the complex is saturated with surfactant [31]. Studies on chemically modified proteins in which the effects of blocking the ionic sites on the binding characteristics have been analyzed [32], together with studies on the binding of anionic surfactants to cationic polypeptides [33] and cationic surfactants to anionic polypeptides [34], demonstrate the electrostatic nature of the initial interaction between protein and surfactant. The initial binding most probably occurs with little change in the tertiary structure of the protein and no change in the secondary structure. The protein–surfactant complexes will have more hydrophobic surface than the protein from which they are formed and at this stage may become insoluble and precipitate. However, increase in the surfactant concentration will result in further binding, and the protein will denature. This unfolding will expose the hydrophobic residues previously buried in the tertiary structure, and surfactant will bind hydrophobically to these newly exposed sites. This further binding almost always results in complexes that are soluble, so that the precipitate formed initially dissolves.

Figure 3 shows typical binding isotherms for hen egg lysozyme on binding sodium *n*-dodecylsulphate (SDS) in acid solution, pH 3.2, 25°C. Lysozyme (molecular mass 14,306) has a single polypeptide chain with an *N*-terminal lysine residue, four disulphide bonds, and 18 cationic residues (11 arginyl, 6 lysyl, and 1 histidyl). At very low free SDS concentrations the binding isotherms are very steep but

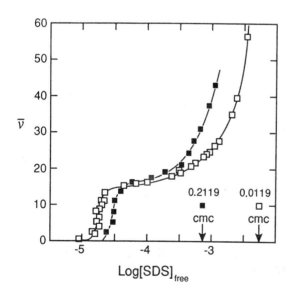

Figure 3 Binding isotherms for the interaction of lysozyme with sodium *n*-dodecylsulphate (SDS) at 25°C, pH 3.2; □, ionic strength 0.0119 M; ■, ionic strength 0.2119 M. (Reproduced from Jones [15] with permission from the Royal Society of Chemistry.)

reach a more gently rising plateau as the ionic interactions with the cationic residues saturate. As the cmc is approached, the binding isotherms rise steeply, characteristic of positive cooperativity. The effect of ionic strength on the binding isotherms is interesting in that, for the initial ionic interaction, increasing the ionic strength weakens the electrostatic interactions between protein and surfactant, so that the binding isotherm is shifted to higher free surfactant concentrations, while it strengthens the hydrophobic interactions, shifting the isotherm at high binding levels to low free surfactant concentration. In this sense the basis for the division of ionic surfactant binding to proteins into specific ionic (polar) and hydrophobic (apolar) interactions is well founded, although in the intermediate region both types of binding may occur together.

Specific ionic binding will of course be significantly affected by changes in pH through changes in the state of ionization of the amino acid side chains. In acid solution, the cationic sites (lysyl pKa ~ 10, histidyl pKa ~ 6, arginyl pKa ~ 12) will be fully protonated, while glutamyl and aspartyl side chains, which will interact repulsively towards anionic surfactants, will be partially protonated, so that acidic conditions are very favorable for the binding of anionic surfactants. In alkaline solution, the glutamyl and aspartyl residues will be fully ionized, and as the pH is increased, the cationic sites will partially lose their positive charge, so that the protein will progressively lose affinity for specific ionic binding of anionic surfactant. It may still however bind hydrophobically, so that we may expect significant changes in the nature of binding isotherms with pH; thus for anionic surfactants at low pH, isotherms may display both specific polar and apolar binding characteristics, while at high pH, only apolar binding will occur.

V. EXPERIMENTAL TECHNIQUES USED IN THE STUDY OF PROTEIN–SURFACTANT INTERACTION

The range of experimental techniques that have been used in the study of protein–surfactant interactions is summarized in Table 1. The most important questions to be answered are (a) the extent of

Table 1 Techniques Used in the Study of Surfactant–
Globular Protein Interaction

Technique	Information obtained [Ref. to examples]
Quantitative equilibrium dialysis	Binding isotherms, Gibbs energy of ligand binding [32,35,36]
Molecular sieve chromatography	Binding levels [37]
Titrimetry	Proton binding in relation to surfactant binding [35,36]
Calorimetry (microcalorimetry and titration calorimetry)	Enthalpy of surfactant binding and protein unfolding [38,39,40,41]
Polyacrylamide gel electrophoresis	Detection of specific complexes [42]
Ultracentrifugation (sedimentation rate and equilibrium)	Sedimentation coefficients of protein–surfactant complexes, subunit dissociation and molecular weights [26,27,43]
Viscometry	Hydrodynamic volume and shape factors, protein unfolding [38,39]
Static and dynamic light scattering	Molecular weights, diffusion coefficients-complex dimensions [44,45,46]
UV difference spectroscopy	Surfactant-induced conformational changes [40]
Neutron scattering	Structure of surfactant–protein complexes [47,48]
X-ray scattering	Structure of protein–surfactant complexes [49]
Enzyme kinetics	Surfactant-induced enzyme denaturation [50,51] or activation [28,29,20]
Surface tension	Surface activity of complexes [52,53]

surfactant binding, (b) the conformational changes induced by binding, and (c) the structure of the complexes that are formed. Of the methods for the measurement of binding, quantitative equilibrium dialysis is the most straightforward and thermodynamically sound. This involves placing a known volume of the protein solution of known concentration in a dialysis membrane bag and equilibrating it against a known volume of surfactant solution. The free concentration of surfactant in equilibrium with the complexes inside the bag is assayed by taking aliquots from outside the bag, since the free concentration is the same on both sides of the dialysis membrane. From the known initial amount of surfactant in the system and the final amount free, the amount bound to the protein can be calculated. The important experimental considerations in this technique are (1) that the dialysis membrane tubing be impermeable to the protein and its complexes but freely permeable to the surfactant; (2) that the ionic strength be sufficiently high to eliminate any Donnan effect, i.e., the development of an unequal concentration of surfactant on either side of the membrane to compensate for the charge on the protein; (3) that the system be in equilibrium at the chosen temperature which may require at least 96 hours for a typical surfactant such as sodium n-dodecylsulphate [32]. If the equilibrium dialysis experiment is carried out over a range of surfactant concentration, the resulting binding isotherm gives a considerable amount of useful information about the binding behavior, including the Gibbs energies of surfactant binding (see below). The binding of surfactant results in complexes that, if the surfactant bound were fully ionized, would have at high binding levels a high surface charge density; but counterion binding reduces the surface charge density. Protons will always be one of the counterions in the case of complexes formed between proteins and anionic surfactants. Proton binding can be determined from titration curves of the protein in the presence of surfactant; such data are relatively difficult to interpret and require the surfactant binding isotherm over a range of pH, but in favorable circumstances the Gibbs energies of the binding of both protons and surfactant to the protein can be evaluated [35,36]. To complete a thermodynamic analysis, the enthalpies of interaction can be measured by titration or microcalorimetry. A very sensitive instrument is required because the energies involved in binding monomeric surfactant below the cmc to a few cm^3 of a protein in dilute solution (<1% w/v) are generally only of the order of 10–50 mJ. The dissociation of protein subunits by surfactant will result in the formation of surfactant–subunit complexes. The dissociation can be followed by both ultracentrifugation sedimentation rate measurements, using absorption or schlieren optics, or by electrophoresis (PAGE). Conformational changes induced by surfactant binding can be followed by viscometry. The intrinsic viscosity $[\eta]$ of a globular protein in its native state depends on its degree of hydration and asymmetry according to the Simha equation [54]

$$[\eta] = \nu(\bar{v}_2 + \delta_1 v_1^0) \tag{1}$$

where ν is the Simha factor, which for a rigid ellipsoid increases with its axial ratio (a/b) (for a sphere $a/b = 1$ and $\nu = 2.5$), $\bar{v}_2$ and v_1^0 are the partial specific and specific volumes of the protein and solvent respectively, and δ_1 is the hydration parameter (grams of water per gram of dry protein). For a roughly spherical (rigid) globular protein ($\bar{v}_2 \sim 0.7$ cm^3 g^{-1}) in aqueous media ($\bar{v}_1^0 \sim 1$ cm^3 g^{-1}), δ_1 is approximately 0.2 g g^{-1}, hence $[\eta] \sim 2.3$ cm^3 g^{-1}. On unfolding, the structure becomes more flexible; the intrinsic viscosity of a flexible protein is given by the Flory-Fox equation [55]

$$[\eta] = \frac{\Phi(\sqrt{r^2})^3}{M} \tag{2}$$

in which Φ is a universal constant, $\sqrt{r^2}$ is the root-mean-square end-to-end distance of the polypeptide chain, and M is the molecular mass. The root-mean-square end-to-end distance is molecular weight dependent and proportional to M raised to a greater power than 1; hence $[\eta]$ for the unfolded protein is molecular weight dependent, in contrast to the native state (Eq. (1)). Surfactant-induced unfolding will thus result in significant increases in the intrinsic viscosity of the protein.

Of the other techniques to follow conformational changes, dynamic light scattering gives the diffusion coefficient, which will decrease with the size of the structure, due to an increase in the frictional coefficient. UV difference spectroscopy can be used to monitor

changes in the environment of specific chromophores that absorb at
~280 nm, such as the aromatic amino acid residues phenylanaline,
tyrosine, and tryptophan, arising from surfactant binding. In the case
of enzymes, the enzymic activity is often a convenient measure of
unfolding, although it is also important to bear in mind that specific
binding at the active site may result in enzyme inhibition before any
appreciable conformational change takes place.

The determination of binding and conformational changes leaves
the question of the detailed structure of complexes unanswered. At
present there is no absolute method for structure determination of
protein–surfactant complexes apart from x-ray diffraction, which has
only been applied to lysozyme with three bound SDS molecules [49].
X-ray diffraction requires a crystal, so in the case of lysozyme cross-
linked triclinic crystals of the protein were soaked in 1.1 M SDS and
then transferred to water or a lower concentration (0.35 M) of SDS
to allow the protein to refold. It was necessary to use cross-linked
crystals to prevent them dissolving when exposed to a high SDS concentration. The resulting denatured–renatured crystals were found to
have three SDS molecules within a structure that was similar but not
identical to that of native lysosyme. Neutron scattering has been applied in a few cases (see Sec. IX), but this is a model-dependent technique.

VI. SURFACE PROPERTIES OF PROTEIN–SURFACTANT SYSTEMS

The interfacial behavior of protein–surfactant complexes is important
in several areas such as the stability of emulsions and foams and the
adsorption of proteins and surfactants from their binary solutions onto
solid surfaces. Of particular interest is the adsorption of the milk
proteins β-lactoglobulin and β-casein at the oil–water interface in the
presence of nonionic surfactants in relation to food emulsions [56–
58] and foam stability [59]. The adsorption of gelatin at the air–water
[52,53,60], oil–water [6], and solid–water [62] interfaces in the presence of surfactants has also been studied. Other studies reported include adsorption from aqueous solutions of lysozyme plus ionic surfactants at solid surfaces [63,64], β-lactoglobulin plus SDS onto

methylated silica [65], fibrinogen plus surfactants on wettability gradient surfaces [66], and the aqueous–air interface [67,68]. Adsorption of proteins and surfactants separately and in combination on solid surfaces with chemical composition gradients have been reviewed by Elwing and Gölander [69].

Adsorption at the aqueous–air interface from binary solutions of proteins and surfactants can be conveniently followed by surface tension measurements in which the protein concentration is kept constant and the surfactant concentration is increased to concentrations in excess of the cmc. Studies of this type were first carried out not with proteins but with polyethylene oxide in the presence of SDS [70], and it was found that plots of surface tension as a function of surfactant concentration showed a number of interesting features in comparison with the surface tension concentration plot in the absence of polymer (Fig. 4). Very similar behavior to that first observed for the polyethylene oxide–SDS system has been found for protein–surfactant systems including bovine serum albumin plus SDS [67], gelatin plus SDS [52], and reduced lysozyme plus hexa (oxyethylene) dodecyl

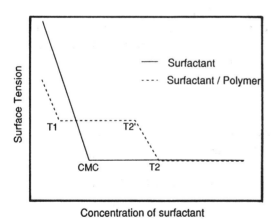

Figure 4 Schematic surface tension vs. surfactant concentration plots in the presence (- - -) and absence (—) of a polymer. T1 indicates the critical concentration of binding. T2′ the saturation concentration, and T2 the formation of surfactant micelles. (Reproduced from Rades et al. [53].)

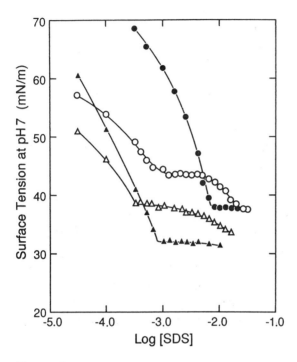

Figure 5   The effect of gelatin on the surface tension of sodium *n*-dodecyl-sulphate (SDS) in water and 0.2 M sodium chloride. ● SDS in water; ○, SDS in 0.5% gelatin; ▲, SDS in 0.2 M sodium chloride; △, SDS in 0.5% gelatin, 0.2 M sodium chloride. (Reproduced from Knox and Parshall [52] with permission from Academic Press.)

ether ($C_{12}E_6$) [71]. Figure 5 shows the surface tension–log [SDS] plots for gelatin solutions in comparison with those for SDS in the absence of protein. Three regions in the surface tension plots can be recognized. At low SDS concentrations the gelatin–SDS systems have a lower surface tension than pure SDS. This surface tension lowering is significantly greater than the sum of the lowerings observed for the two pure components, from which it can be inferred that gelatin–SDS complexes are more surface active than SDS. The surface tension lowering comes to an abrupt stop at an SDS concentration of 1×10^{-3} M, and a plateau is formed up to an SDS concentration of approximately 5×10^{-3} M, after which the surface tension falls to that of pure SDS. The plateau region is suggested to arise as the complexes, enriched in SDS, are adsorbed. The random-coil conformation of gelatin prevents the close packing in the interface of the dodecyl sulphate groups bound to cationic sites on the gelatin. The fall in surface tension above 5×10^{-3} M SDS is believed to arise as the complexes become saturated with surfactant molecules bound hydrophobically by their alkyl groups with a constant increase in solubility and a decrease in surface activity. The surface tension behavior at pH 7 has been correlated with the solubility of the gelatin–SDS complexes at pH 4.1 (Fig. 6) [52]. The fraction of gelatin that precipitates from the solutions at pH 4.1 passes through a maximum

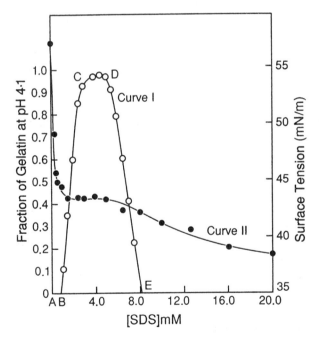

Figure 6 Comparison of the precipitation at pH 4.1 and the surface tension at pH 7 for sodium *n*-dodecylsulphate (SDS) in 0.5% gelatin. Curve I, fraction of gelatin precipitated vs. SDS concentration; curve II, surface tension vs. SDS concentration. (Reproduced from Knox and Parshall [52] with permission from Academic Press.)

in the region of the surface tension plateau at pH 7.0. At low SDS concentrations (up to 0.5×10^{-3} M) the complexes are soluble at pH 4.1 (region AB in Fig. 6). Between 1×10^{-3} M and 3×10^{-3}M (region BC) the complexes contain 0.13 to 0.17 g SDS per gram of gelatin precipitate. The maximum precipitation region (3×10^{-3} to 5×10^{-3} SDS, region CD) occurs as the complex compositions increase in SDS content from 0.18 to 0.25 g SDS per gram of gelatin. At higher SDS concentrations ($>5 \times 10^{-3}$ M) the complexes become more soluble, and at 8×10^{-3} M SDS no precipitate is formed.

The situation with regard to complex formation by nonionic surfactants is not clearly defined. Knox and Parshall [52] found that the surface tension–log(concentration) curves of both Aerosol OT (sodium di (iso-octyl) succin-1-sulphonate) in 0.2 M sodium chloride and Triton X100 in water in the presence of gelatin were identical to those for the surfactant alone, although in water the Aerosol OT plus gelatin system gave a surface tension curve indicative of the formation of surface-active complexes. Reduced lysozyme with the non-ionic $C_{12}E_6$ gave a surface tension curve characteristic of complex formation, which was also detected by equilibrium dialysis [71]. In contrast, Rades et al. [53] found no evidence for complex formation between gelatin and the nonionics $C_{16}E_{15}$, $C_{16}E_{20}$, and $C_{16}E_{25}$ from surface tension measurements.

The behavior of protein–surfactant systems at the oil–water interface is similar to that at the air–water interface. For example, the interfacial tension–log[SDS] plot in the presence of sodium caseinate has very similar characteristics to that for gelatin plus SDS (Fig. 5) in showing an interfacial tension below that of SDS in the presence of caseinate at a low concentration followed by a plateau as the SDS concentration is raised [72]. As for the air–water interface, complexes are initially adsorbed and then displaced as the SDS-to-protein concentration ratio is increased. In the case of protein–nonionic surfactant systems at the oil–water interface, e.g., β-casein plus $C_{12}E_8$, complexation is either very weak or nonexistent, and the protein and surfactant complete for adsorption sites at the interface [73]. If the protein concentration is constant and the surfactant concentration is increased, the β-casein, which in the absence of surfactant adsorbs at the interface, is progressively displaced until the interface contains only surfactant. Such behavior is important in food emulsions [19], where diplacement of adsorbed protein by nonionic emulsifiers, particularly if they are oil soluble, can result in coalescence of oil-in-water emulsion droplets.

Adsorption for protein–surfactant mixtures at solid–water interfaces is controlled by a number of factors and involves competition between the protein and the surfactant for the solid surface. Surfactants generally adsorb reversibly on solids and hence can be removed by dilution of the aqueous phase (rinsing); in contrast, some proteins undergo irreversible adsorption. An example of the complexity of events on adsorption from a protein–surfactant mixture as a function of dilution of the aqueous phase is shown in Fig. 7 for the adsorption

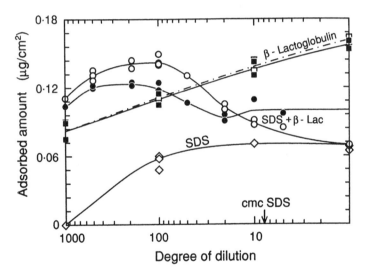

Figure 7 The adsorption of sodium n-dodecylsulphate (SDS), β-lactoglobulin, and β-lactoglobulin from a mixture of 0.1% w/v protein plus 0.5% w/v SDS onto methylated silica as a function of degree of dilution. $\Diamond$, pure SDS; $\Box$, $\blacksquare$ pure β-lactoglobulin; $\bigcirc$, $\bullet$ SDS + β lactoglobulin. The open symbols are adsorption after 30 minutes, and the closed symbols are adsorption after 30 minutes of rinsing. Solvent 0.01 M phosphate buffer, pH 7 plus 0.15 M sodium chloride. (Reproduced from Wahlgren and Arnebrant [65] with permission from Academic Press.)

from β-lactoglobulin–SDS mixtures onto a methylated silica surface [65]. This surface adsorbs β-lactoglobulin irreversibly; it cannot be washed off, and adsorption increases almost linearly with increasing concentration. In contrast, SDS adsorbs reversibly and the surface saturates at the cmc. Adsorption of β-lactoglobulin from its mixtures with SDS, in which it is complexed, passes through a maximum as the solutions are diluted. At high concentrations the amount of β-lactoglobulin adsorbed is less than for the pure protein, the surfactant-to-protein ratio in the complexes will be high, and they will be less surface active than the pure protein. On dilution SDS will partially dissociate from the complexes and they will become more surface active than the pure protein. If the surface is rinsed, the ratio of surfactant to protein will decrease as the surfactant more readily desorbs than the protein, and furthermore dilution of the complexes will lead to surfactant dissociation. As a consequence, the rinsed surface becomes richer in protein relative to adsorption from the protein–surfactant mixture at high concentrations and poorer in protein relative to the mixture at low concentrations (high dilution).

Wahlgren and Arnebrant have identified several different types of adsorption/displacement mechanisms at solid surfaces [63] as shown in Fig. 8. In these models the protein-covered surface is treated with surfactant solutions above the cmc. In the first case (A), the addition of surfactant results in the formation of soluble complexes that desorb leaving a clean surface; examples of this mechanism are β-lactoglobulin and lysozyme adsorbed on silica treated with SDS. In the second class (B), the surfactant replaces the protein on the surface; this is found for β-lactoglobulin adsorbed on silica when treated with cetyltrimethylammonium bromide (CTAB) or β-lactoglobulin adsorbed on methylated silica when treated with CTAB or SDS. The third class (C) involves the reversible interaction of the surfactant with the adsorbed protein. Here protein adsorption is stronger than surfactant adsorption, and on rising the protein is left on the surface. Examples of this are β-lactoglobulin adsorbed on nickel oxide plus SDS and β-lactoglobulin or lysozyme adsorbed on chromium oxide plus CTAB. The final class (D) is similar to C but results in partial removal of the protein from the surface. The behavior of protein–surfactant–solid surface systems is thus dependent on a range of factors, and

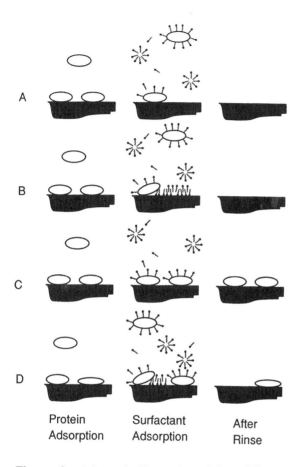

Protein   Surfactant   After
Adsorption   Adsorption   Rinse

Figure 8 Schematic illustration of four different adsorption/displacement models proposed by Wahlgren and Arnebrant [63] for protein and surfactant adsorption to solid surfaces. The three diagrams for each model show protein adsorption, surfactant addition, and state after rinsing. Figure A represents the case where surfactant binds to the protein and the protein–surfactant complex desorbs. Figure B represents protein displacement by the surfactant. Figure C represents reversible adsorption of the surfactant by the protein. Figure D represents reversible adsorption by the surfactant resulting in partial desorption of the protein. The figures relate to a hydrophilic surface; at a hydrophobic surface the orientation of the surfactant molecules with respect to the surface will be different. (Reproduced from [63] with permission from Academic Press.)

each system must be individually analyzed to establish its characteristics.

VII. THERMODYNAMICS OF PROTEIN– SURFACTANT INTERACTIONS

The thermodynamic parameters characterizing protein–surfactant interactions can be obtained by combining the analysis of binding isotherms, most commonly obtained by equilibrium dialysis, with microcalorimetric measurements of the enthalpies of interaction. In principle, measurements of the binding isotherms over a range of temperature can also give the enthalpies of interaction, but in general the available temperature range is restricted by the thermal stability of the protein, and binding isotherms of very high precision are required to obtain reliable enthalpy data by this method. Thermodynamic data for many of the most commonly available globular proteins on interaction with surfactants have been reported as summarized in Table 2. In very general terms, the Gibbs energy of binding per mole of surfactant ($\Delta G_{\bar{\nu}}$, see below) is usually of the order of -20 to -25 kJ mol^{-1}, while the enthalpy per mole of surfactant bound ($\Delta H_{\bar{\nu}}$), is of the order of 0 to $\pm$ 10 kJ mol^{-1}, so that the positive entropic contribution to the interaction ($T\Delta S_{\bar{\nu}}$) is a major driving force in complex formation. Thus, like micelle formation, hydrophobic interactions play a major role.

For a rigorous interpretation of complexation it is important to appreciate that the formation of protein–surfactant complexes is an example of multiple equilibria between monomeric surfactant and complexes having different numbers of bound surfactant molecules. These equilibria, which we assume are reversible (see Sec. VIII), can be represented by a set of equations in which P is the protein and S the surfactant:

$$P \ + S \rightleftharpoons PS_1$$
$$PS_1 + S \rightleftharpoons PS_2 \qquad (2a)$$
$$PS_2 + S \rightleftharpoons PS_3$$

or more generally

$$PS_{i-1} + S \rightleftharpoons PS_i \qquad (3)$$

Table 2 Some Thermodynamic Studies on Protein–Surfactant Interactions

Protein	Surfactants	Conditions	Ref.
Ribonuclease A	SDS	pH 7.1, 25°C	40
Bovine serum albumin	SDS	pH 7.0, I = 0.005, 18°–32°C	38
Ovalbumin	SDS	pH 7.0, I = 0.005, 18°–32°C	38
Ribonuclease A	SDS	pH 7.0, I = 0.005, 18°–32°C	38
Bovine serum albumin	Potassium n-dodecylsulphate, DTAB	pH 7.0, I = 0.005, 25°C	39
β-Lactoglobulin	SDS	pH 3.5–7.0, I = 0.01, 25°C	74
Trypsin	SDS	pH 3.5, 5.5, I = 0.01, 25°C	50
Lysozyme	Sodium n-octyl, n-decyl, n-dodecylsulphates	pH 3.2, I = 0.0088, pH 9.0, I = 0.004, 25°C	32,36,75,76
Lysozyme	SDS	pH 3.2, I = 0.0119–0.2119, 25°C	77
Glucose oxidases	SDS	pH 4.3–6.0, I = 0.0055–0.0155, 25°C	26
Catalases	SDS	pH 3.2–10.0, I = 0.0119–0.2, 25°C	27,35,78,79,80,81,89
Myosins	SDS	pH 6.4, I = 0.6, 25°C	82
Range of globular proteins	n-Octylglucoside	pH 6.4, I = 0.132, 25°C	83
Catalases	DTAB	pH 3.2–10.0, I = 0.0119–0.0318, 25°C	84
Insulin	Sodium n-octyl, n-decyl, n-undecyl, n-dodecyl sulphates	pH 3.2, I = 0.025, pH 10.0, I = 0.125, 25°C	85,86,87
Glucose oxidase	C_8–C_{12} n-alkylsulphates, C_8, C_{10}, C_{12} TABS	pH 10 (TABs), pH 3.2 (sulphates), 25°C	88
Histones	SDS	pH 3.2–10.0, I = 0.0119–0.0318, 25°C	90–94

The equilibrium constants, assuming ideality (activity coefficient of unity), for the first three steps are given by

$$K_1 = \frac{[PS_1]}{[P][S]} \qquad K_2 = \frac{[PS_2]}{[PS_1][S]} \qquad K_3 = \frac{[PS_3]}{[PS_2][S]} \tag{4}$$

Eliminating $[PS_1]$ and $[PS_2]$ by substitution of $[PS_1]$ in K_2 and $[PS_2]$ in K_3 gives

$$K_1 K_2 K_3 = \frac{[PS_3]}{[P][S]^3} \tag{5}$$

Hence for a total of n steps

$$K_1 K_2 K_3 \cdots K_n = \frac{[PS_n]}{[P][S]^n} \tag{6}$$

If the equilibrium constants are identical for the n equilibria ($K_1 = K_2 = K_3 = \cdots = K_n = K$) then

$$K^n = \frac{[PS_n]}{[P][S]^n} \tag{7}$$

The average number of surfactant molecules bound per protein molecule ($\bar{\nu}$) will then be given by the concentration of bound surfactant divided by the total concentration of protein.

$$\bar{\nu} = \frac{{}^n\Sigma_{i=1} \, i[PS_i]}{[P] + {}^n\Sigma_{i=1} \, [PS_i]} \tag{8}$$

hence

$$\bar{\nu} = \frac{[P]\{K_1[S] + 2K_1K_2[S]^2 - n(K_1K_2 - K_n)[S]^n\}}{[P] + [P]\{K_1[S] + K_1K_2[S]^2 - (K_1K_2 - K_n)[S]^n\}} \tag{9}$$

While Eq. (9) gives $\bar{\nu}$ as a function of surfactant concentration, without values of the individual binding constants it is of limited practical use. If however equilibrium constants are all equal, then from Eq. (7) it follows that

$$\bar{\nu} = \frac{n[PS_n]}{[P] + [PS_n]} = \frac{n(K[S])^n}{1 + (K[S])^n} \tag{10}$$

In general, since the equilibrium constants will not be equal, Eq. (10) would not be expected to hold, but to preserve the simple form of the equation, Hill introduced a cooperativity coefficient n_H to replace n, giving [95]

$$\bar{\nu} = \frac{n([K[S])^{n}_{H}}{1 + (K[S])^{n}_{H}} \tag{11}$$

For a positively cooperative process (i.e., binding of ligands enhances further binding), $n_H > 1$, while for a negatively cooperative process (i.e., binding of ligands inhibits further binding) $n_H < 1$.

An alternative approach to binding is based on the Scatchard equation [96]. If a protein has n independent and identical binding sites with intrinsic binding constants K and a fraction Θ of these are occupied at a given surfactant concentration [S], then a simple kinetic argument, in which the rate of binding is proportional to [S] times the fraction of vacant sites (1-Θ) and is equated to the rate of dissociation from the occupied sites proportional to Θ, gives

$$\Theta = K[S](1\text{-}\Theta) \tag{12}$$

where K is a binding constant; since $\Theta = \bar{\nu}/n$ it follows that

$$\frac{\bar{\nu}}{n} = K[S]\left(1 - \frac{\bar{\nu}}{n}\right) \tag{13}$$

hence

$$\bar{\nu} = \frac{nK[S]}{1 + K[S]} \tag{14}$$

Equation (14) is identical to the Hill equation (11) when the Hill coefficient is unity. It follows from the Scatchard equation that

$$\frac{\bar{\nu}}{[S]} = K(n - \bar{\nu}) \tag{15}$$

so that the plot of $\bar{\nu}/[S]$ is a linear function of $\bar{\nu}$ with slope K and intercept on the abscissa of n.

It is informative to see how the shape of binding isotherms and their corresponding Scatchard plots based on Eq. (15) depend on co-

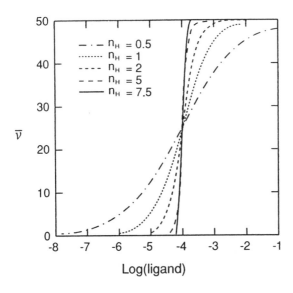

Figure 9   Theoretical binding isotherms ($\bar{v}$ vs. log[ligand]) calculated from the Hill equation for a macromolecule (protein) with 50 binding sites (intrinsic binding constant 10^4) for a range of Hill coefficients n_H. (Reproduced from Jones [15] with permission from the Royal Society of Chemistry.)

operativity (n_H). Figure 9 shows a series of binding isotherms generated from the Hill equation for a hypothetical molecule (e.g., protein) with 50 binding sites with intrinsic binding constant 10^4, for various Hill coefficients from 0.5 (negative cooperativity) to 7.5 (highly positive cooperativity) [97]. The curves are sigmoidal and increase in steepness with increasing n_H. There is no way of distinguishing between negative and positive cooperativity from inspection, since the curves are qualitatively similar. In contrast, the Scatchard plots for the same systems show dramatically changing characteristics (Fig. 10) that are diagnostic of the type of cooperativity [98]. For Hill coefficients less than unity, the Scatchard plots have negative slopes decreasing with increasing v that intercept the abscissa at n (50); for $n_H = 1$ the plot is linear, as required by Eq. (15), whereas for $n_H > 1$ the plots pass through maxima, the maximum point becoming

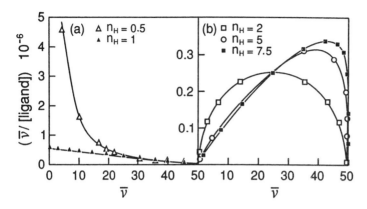

Figure 10 Scatchard plots ($\bar{\nu}/$[ligand] vs. $\bar{\nu}$) for the isotherms of Figure 9. (a) $n_H = 0.5$ (negative cooperativity) and $n_H = 1$ (independent noninteracting sites). (b) $n_H = 2$ to 7.5 (positive cooperativity). (Reproduced from Jones [15] with permission from the Royal Society of Chemistry.)

asymmetrically displaced to higher values of $\bar{\nu}$ with increasing n_H. The misuse of Scatchard plots for $n_H < 1$ has been the subject of much controversy [99–101]. The problem relates primarily to extrapolation of the roughly linear part of the plot for $n_H < 1$ to the abscissa to obtain an intercept that is taken as the total number of binding sites. For example, in Fig. 10a, for $n_H = 0.5$ the steep part of the plot could be extrapolated to give n ~ 12–13, but such an extrapolation has no significance in this context.

Examples of both negative and positive cooperativity are found when the Scatchard analysis is applied to protein–surfactant interactions. Figure 11 shows the Scatchard plot for the binding of SDS to bovine catalase (molecular mass 245,000) in acid solution at pH 3.2. The curve is diagnostic of negative cooperativity and can be extrapolated to give values of n of 343 ± 6, which is close to the number of cationic amino acid residues in the catalase molecule, 331 (112 lysyl, 86 histidyl, and 133 arginyl). The flatter part of the curve corresponds to further binding, largely of a hydrophobic type. In contrast, at pH 6.4 a typical positively cooperative Scatchard plot is found corresponding to a Hill coefficient of 2.61 ± 0.07 (Fig. 12).

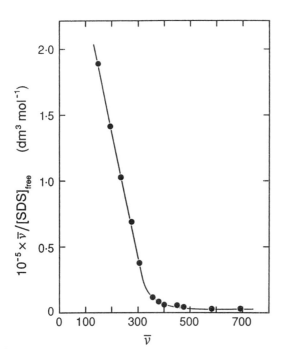

Figure 11   Scatchard plot for sodium n-dodecyl sulphate (SDS) binding to bovine catalase at 25°C, pH 3.2. (Adapted from Ref. 35.)

Numerous other protein–surfactant systems when subjected to Scatchard analysis give curves that extrapolate to a number of specific binding sites very close to the number of cationic amino acid residues in the protein. Furthermore, chemical modification of the cationic sites, e.g., in the case of lysyl residues by acetylation, shifts the Scatchard plots to give lower values of n [32]. Despite these interesting observations, Scatchard analysis should be treated with a degree of caution; it is not entirely clear why such a good correspondence between n and the number of cationic sites is obtained, since the binding sites must only approximate to independence and are certainly not chemically identical.

A more rigorous analysis of binding isotherms can be made by the application of the binding potential concept proposed by Wyman

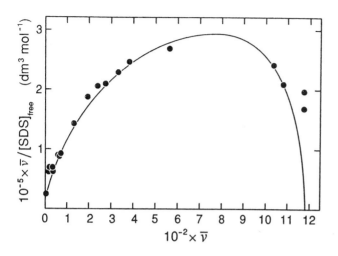

Figure 12 Scatchard plot for sodium *n*-dodecylsulphate (SDS) binding to bovine catalase, 25°C, pH 6.4. The solid line was fitted using the Hill equation for a total of 1190 binding sites per catalase molecule (1.4 g SDS per g catalase) and corresponds to an intrinsic binding constant of 479 ± 6 dm^{-3} mol^{-1} and $n_H = 2.62 \pm 0.07$. (Adapted from Ref. 35.)

[102]. Wyman recognized that there should be a function, which he called the binding potential $\pi(P, T, \mu_i, \mu_j \ldots)$, relating the extent of binding ($\bar{\nu}$) to the chemical potential of any ligand (μ) such that

$$\bar{\nu} = \left(\frac{\partial \pi}{\partial \mu} \right)_{P, T, \mu_i, \mu_j \ldots} \tag{16}$$

If the chemical potential of a surfactant ligand (μ_s) is given by the ideal solution expression

$$\mu_s = \mu^o_s + RT \ln[S] \tag{17}$$

where μ^o_s is the standard chemical potential, here at unit molarity, then it follows from Eq. (16) that

$$\bar{\nu} = \frac{1}{RT} \left(\frac{\partial \pi}{\partial \ln[S]} \right)_{P, T, \mu_i, \mu_j \ldots} \tag{18}$$

Hence the binding potential will be given by

$$\pi = 2.303 \ RT \int_{0}^{\bar{\nu}} \bar{\nu} \ d \ \log[S] \tag{19}$$

It follows from Eq. (19) that the binding potential can be calculated by integrating under the binding isotherm ($\bar{\nu}$ vs. $\log[S]$) from $\bar{\nu} = 0$ to any desired value of $\bar{\nu}$. If $\bar{\nu}$ is expressed in the form of Eq. (10) then it can be shown [14,75] that

$$\pi = 2.303 \ RT \ \log(1 + K^n[S]^n) \tag{20}$$

If it is assumed that for any given free surfactant concentration for an extent of binding $\bar{\nu}$, corresponding to a complex PS_ν, then

$$\pi = 2.303 \ RT \ \log(1 + K_{app} \ [S]^\nu) \tag{21}$$

Since π can be evaluated by integration under the binding isotherm at any desired value of $\bar{\nu}$ corresponding to a particular free surfactant concentration, it is possible to determine a range of apparent binding constants K_{app} as binding proceeds. The Gibbs energy of binding per ligand bound can then be determined from

$$\Delta G_{\bar{\nu}} = -\frac{RT}{\bar{\nu}} \ln K_{app} \tag{22}$$

The calculation of $\Delta G_{\bar{\nu}}$ from Eq. (22) does not take into account the statistical contribution to the entropy and hence Gibbs energy of binding. This arises because ligands can be arranged on the binding sites in a number of different ways, making a positive contribution to the entropy of binding per ligand bound ($\Delta S_{\bar{\nu},stat}$). If the protein has n indistinguishable binding sites, the number of arrangements $\Omega_{n,i}$ of ligands on the n sites is given by

$$\Omega_{n,i} = \frac{n!}{(n-i)!\,i!} \tag{23}$$

and the contribution to the entropy of binding will be given by the Boltzmann equation ($R \ln \Omega_{n,i}$). The Gibbs energy per ligand bound from this source will thus be

$$\Delta G_{\bar{\nu},\text{stat}} = -\frac{RT}{i} \ln \Omega_{n,i} \qquad (24)$$

This contribution to $\Delta G_{\bar{\nu}}$ is greatest for the first ligand bound and decreases as the sites are filled and there are fewer possible arrangements of the ligands on the binding sites. The effect of this statistical correction on the thermodynamic parameters is illustrated in Fig. 13 for the binding of n-dodecyltrimethylammonium bromide (DTAB) to glucose oxidase (25°C, pH 10.). Here in calculating $\Delta G_{\bar{\nu},\text{stat}}$ it was assumed that glucose oxidase has a total of 1700 binding sites for DTAB. For low binding levels both $\Delta G_{\bar{\nu}}$ and $T\Delta S_{\bar{\nu}}$ are significantly reduced when this statistical contribution is subtracted from the parameters. There are, however, two problems in making statistical cor-

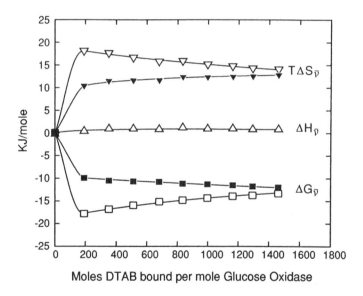

Moles DTAB bound per mole Glucose Oxidase

Figure 13 Thermodynamic parameters for the interaction between glucose oxidase and n-dodecyltrimethylammonium bromide (DTAB) as a function of the number of moles of DTAB bound per glucose oxidase molecule (ν). $\Delta G_{\bar{\nu}}$ ($\square$), $\Delta H_{\bar{\nu}}$ ($\triangle$), and $\Delta S_{\bar{\nu}}$ ($\triangledown$) are the Gibbs energy, enthalpy, and entropy per mole of DTAB bound. The symbols ■ and ▼ correspond to $\Delta G_{\bar{\nu}}$ and $T\Delta S_{\bar{\nu}}$ corrected for statistical effects. (Data taken from Ref. 88.)

rections to $\Delta G_{\bar{\nu}}$ and $\Delta S_{\bar{\nu}}$: first, the total number of binding sites needs to be known, and this is not always easily found from the binding isotherms; second, binding initially occurs to the native protein (specific binding) before unfolding is promoted. After initial specific binding, the protein unfolds, exposing a larger number of potential binding sites to which the surfactant binds nonspecifically and finally cooperatively. Thus the statistical corrections change on unfolding. Attempts have been made to take these effects into account in the case of the binding of SDS to lysozyme, a system that has been studied in some detail [97].

Thermodynamic parameters (without statistical corrections) for the interaction of surfactant with a range of globular proteins are shown in Table 3. For these systems a number of generalizations can be made. Firstly, binding isotherms for ribonuclease A, lysozyme, ovalbumin, glucose oxidase, and bovine catalase all show clear evidence of specific binding of SDS to the cationic sites on the surface of the native protein. Thus the numbers of cationic amino acid residues for ribonuclease and lysozyme are 18, for ovalbumin 42, for glucose oxidase 119, and for bovine catalase 343. For these globular proteins the average values $\Delta G_{\bar{\nu}}$ is of the order of -27 kJ mol^{-1}. $\Delta G_{\bar{\nu}}$ for nonspecific binding is generally of lower energy decreasing to of the order of -17 kJ mol^{-1} at high binding levels. The binding of the nonionic surfactant OBG (n-octyl-β-D-glucopyranoside) to globular proteins is controversial. Cordoba et al. [83] found clear evidence for binding using equilibrium dialysis and showed that for a wide range of globular proteins the maximum binding levels corresponded approximately to the adsorption of a monolayer of OBG on the native surface of the protein; but using molecular sieve chromatography Lundahl et al. [37] failed to detect any binding of OBG to globular proteins. Analysis of the equilibrium dialysis data gave values of $\Delta G_{\bar{\nu}}$ of the order of -10 kJ mol^{-1}, appreciably smaller than for the binding of ionic surfactants.

The enthalpies of interaction ($\Delta H_{\bar{\nu}}$) of surfactants with proteins are small relative to $T\Delta S_{\bar{\nu}}$ and of variable sign. Several processes contribute to the overall enthalpy change. The enthalpy per mole of protein, defined as ΔH_{obs}, where $\Delta H_{obs} = \bar{\nu}\Delta H_{\nu}$, contains contributions from the enthalpy of interaction with specific ionic sites (ΔH_{ion}), nonspe-

Table 3 Some Thermodynamic Parameters for the Formation of Protein–Surfactant Complexes

Protein–surfactant	Conditions	$\bar{\nu}$ (Binding type)	$\Delta G_{\bar{\nu}}$	$\Delta H_{\bar{\nu}}$	$T\Delta S_{\bar{\nu}}$	Ref.
Ribonuclease A–SDS	pH 7, 25°C, I=0.005	18 (specific)	−27.9	−1.27	26.6	38
Ribonuclease A–SDS	pH 3.2, 25°C	18 (specific)	−29.6	+1.26	30.9	103
Ribonuclease A–OBG*	pH 6.4, 25°C, I=0.132	100 (nonspecific)	−11.2	0.311	11.5	83
Lysozymes–SDS	pH 3.2, I=0.0119–0.2119	16–17 (specific)	−26.8–25.7	8.4–10.9	18.5–14.6	77
Lysozyme–OBG	pH 6.4, 25°C, I=0.132	130 (nonspecific)	−10.2	0.783	11.0	83
Ovalbumin–SDS	pH 7, 25°C, I=0.005	42 (specific)	−30.0	0	30.0	38
Bovine serum albumin–SDS	pH 7, 25°C, I=0.005	98 (specific)	−24.8	−6.95	17.9	38
Bovine serum albumin–OBG	pH 6.4, 25°C, I=0.132	550 (nonspecific)	−9.7	0.668	10.4	83
Glucose oxidase–SDS	pH 3.7, 25°C	120 (specific)	−21.2	−3.29	17.9	26
Glucose oxidase–SDS	pH 3.2, 25°C	575 (nonspecific)	−17.0	−3.34	13.7	88
Glucose oxidase–DTAB	pH 10.0, 25°C	510 (nonspecific)	−15.8	0.845	16.7	88
Bovine catalase–SDS	pH 3.2, 25°C	331 (specific)	−28.4	−8.36	20.0	35
Bovine catalase–SDS	pH 6.4, 25°C, I=0.0119	500 (nonspecific)	−21.5	−3.5	18.0	80
Bovine catalase–OBG	pH 6.4, 25°C, I=0.132	1900 (nonspecific)	−9.7	0.668	10.4	83
Aspergillus niger catalase–SDS	pH 6.4, 25°, I=0.0119	500 (nonspecific)	−16.0	−2.4	13.6	80
Insulin–SDS	pH 3.2, 25°C, I=0.025	20 (nonspecific)	−17.0	−3.07	13.9	85
Insulin–SDS	pH 10.0, 25°C, I=0.125	20 (nonspecific)	−16.7	−0.80	15.9	85

*OBG = n-octyl-β-D-glucopyranoside.

cific hydrophobic interactions (ΔH_{hc}), and the enthalpy of protein unfolding (ΔH_u). Thus

$$\Delta H_{obs} = \Delta H_{ion} + \Delta H_{hc} + \Delta H_u \qquad (25)$$

The ionic contributions can be estimated from model studies on the interactions of surfactants with polypeptides. Thus for the n-alkylsulphates the enthalpies of interaction with lysyl, histidyl, and arginyl residues have been measured from their interactions with polylysine, polyhistidine, and polyarginine [33]. From the numbers of cationic residues in a particular protein it is possible to estimate ΔH_{ion} from the expression

$$\Delta H_{ion} = \Sigma_r \, n_r \, \Delta H_r \qquad (26)$$

where n_r is the number of moles of residue r per mole of protein and ΔH_r is the enthalpy of interaction per mole of residue r. The estima-

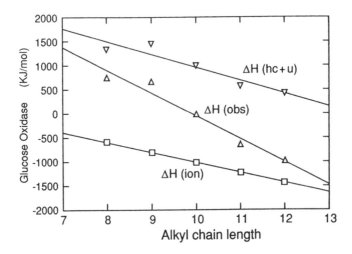

Figure 14 Contributions to the observed enthalpy of interaction (ΔH_{obs}, $\triangle$) between glucose oxidase and n-alkylsulphates at 25°C, pH 3.2 at a relative concentration c/cmc = 0.5, as a function of alkyl chain length. ΔH_{ion} ($\square$) is the contribution from ionic interactions, and $\Delta H_{hc} + \Delta H_u$ (∇) are contributions from hydrophobic binding plus protein unfolding. (Reproduced from Ref. 88 with permission of the publishers, Butterworth-Heinemann Ltd.)

tion and separation of the other two contributions $\Delta H_h + \Delta H_u$ is more problematic. For the particular case of the interaction between a homologous series of n-alkylsulphates with glucose oxidase in acid solution, pH 3.2 [88]; the contributions to ΔH_{obs} are shown as a function of alkyl chain length in Fig. 14. (N.B. The data relate to surfactant concentrations at a concentration of 0.5 cmc where the specific ionic interactions are saturated.) The ionic contributions are exothermic; they increase with alkyl chain length because they include not only a contribution from the cationic residue–sulphate group interaction but also a small contribution from the alkyl chains. Subtracting ΔH_{ion} from ΔH_{obs} gives $\Delta H_{hc} + \Delta H_u$. If $\Delta H_{hc} + \Delta H_u$ is extrapolated to zero alkyl chain length where it is reasonable to assume $\Delta H_{hc} = 0$, a value of 3610 ± 560 kJ mol^{-1} is obtained for ΔH_u. This figure is comparable to the enthalpies of unfolding of many globular proteins when converted to unit mass (23 ± 4 Jg^{-1}) [104].

VIII. THE REVERSIBILITY OF PROTEIN–SURFACTANT COMPLEX FORMATION

In general the complexation of proteins by surfactants is reversible in that the surfactant can be dialyzed out of the system. However the protein may not return to its native state if extensive denaturation has occurred. If the protein has not been reduced by use of mercaptoethanol as in SDS-PAGE then the probability that the protein can be returned to its native state will be much greater. Under certain circumstances small numbers of surfactant molecules can become trapped in the native state structure without appreciable disruption of the structure [49].

The dissociation of protein–surfactant complexes can be induced by the presence of another surfactant with a lower hydrophobic–hydrophilic balance (higher cmc) due to the formation of mixed micelles [105]. This effect is illustrated in Fig. 15 for complex formation between insulin and SDS in the presence of a constant concentration of sodium n-decylsulphate (SDeS). In a binary mixture of similar surfactants (SDS + SDeS) that behave as an ideal mixture in the micelles and in the solution, the cmc (C_{12}^*) is given by [106,107]

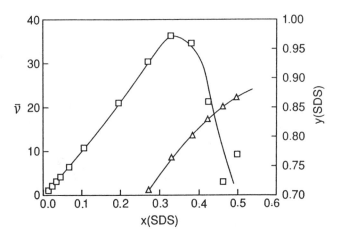

Figure 15   Binding (left-hand axis) of sodium n-dodecylsulphate (SDS) to insulin (pH 3.2. $I = 0.0065$ M) at 25°C as a function of mole fraction of SDS ($\square$) in the presence of a constant concentration (10 mM) of sodium n-decylsulphate.   The right-hand axis shows the mole fraction of SDS in mixed micelles as a function of mole fraction of SDS in the system ($\triangle$). Reproduced from Ref. 105 with permission from the Royal Society of Chemistry.)

$$C_{12}{}^{*} = \frac{C_1 C_2}{x_1 C_1 + x_2 C_2} \tag{27}$$

where x_1 and x_2 are the mole fractions of the two surfactants in the mixed solute ($x_1 + x_2 = 1$) with cmcs C_1 and C_2 in the pure states respectively. The composition of the mixed micelles in terms of mole fractions of the two components (y_1 and y_2, where $y_1 + y_2 = 1$) is given by

$$y_1 = x_1 \frac{C^{*}{}_{12}}{C_1} \qquad y_2 = x_2 \frac{C^{*}{}_{12}}{C_2} \tag{28}$$

In contrast to a single surfactant system the monomer concentrations of the two components in the mixed system do not remain constant above the cmc ($C_{12}{}^{*}$). The monomer concentration of the more hydrophobic component decreases while that of the less hydrophobic component increases, and consequently the micelles became richer in the more hydrophobic component. Moreover, for a system in which the concentration of the more hydrophobic component is increased, at a constant concentration of the less hydrophobic component, the cmc decreases. Thus the consequence of the depression of the cmc of the system, by increasing the total SDS concentration on the binding of SDS to protein (insulin) at constant SDeS concentration, is that once the total surfactant concentration exceeds the cmc the micelles became enriched in SDS at the expense of the SDS–insulin complexes. The binding isotherm thus goes through a maximum as shown in Fig. 15. This process of complex dissociation as a consequence of competition for the more hydrophobic surfactant between complexes and mixed micelles has been termed "retrograde dissociation" [105]. It has also been observed for anionic–nonionic mixed micelle systems, although in such cases the analysis is more complex due to the nonideal behavior of the solutions [105].

Retrograde dissociation clearly demonstrates the reversible nature of protein–surfactant interaction. It may also be noted that maxima in protein–surfactant adsorption isotherms have also been observed in lysozyme–SDS systems under certain conditions [108] as the surfactant cmc is approached; these can arise as a consequence of the surfactant solution activity passing through a maximum in the region of the cmc [109].

IX. THE STRUCTURE OF PROTEIN–SURFACTANT COMPLEXES

The structure of protein-surfactant complexes is perhaps their most important property, but it has proved to be very difficult to determine unambiguously, particularly for globular proteins with their disulphide linkages intact. The situation is a little clearer for reduced proteins although here structures are less rigid. The success of the SDS-PAGE technique, which depends on the existence of a uniform charge per unit length for the saturated SDS–polypeptide complexes, suggests that the complexes must have an extended conformation with surfactant molecules bound approximately uniformly along their length. At the other extreme, when only a few molecules of surfactant are bound to an oxidized (disulphide bonds intact) protein, the protein structure may be very little different from that of the native state. In the case of lysozyme–SDS complexes with a few bound SDS molecules, this has been largely confirmed from the x-ray crystallographic studies [49]. To obtain crystallographic data on the lysozyme–SDS complexes it was necessary to cross-link triclinic lysozyme crystals and soak them in SDS (1.1 M) before transferring them to water or a lower concentration of SDS solution (0.35 M) to allow the protein to refold. Examination of the resulting "denatured–renatured" crystals by x-ray diffraction revealed three SDS molecules in the renatured structure, which was similar, but not identical, to native state crystals. The agreement between the structure factors of the renatured crystals and the native cross-linked crystals was 17% (for renaturation in water) and 19% (for renaturation in 0.35 M SDS), and the minimum spacings in the x-ray pattern of renatured and native crystals were 2.9 Å and 1.1 Å respectively. The need to cross-link the crystals detracts somewhat from the significance of the results, although there were only a few cross-links and these were highly flexible. The results clearly showed the location of the SDS molecules with the surfactant head groups forming a salt bridge with positively charged amino acid residues and the hydrocarbon chains making hydrophobic contact with the tertiary structure. Specifically, of the three SDS molecules bound in the renatured crystals, one formed a salt bridge with the terminal lysine with its alkyl chain penetrating deep into the hydrophobic core of the tertiary structure, while the other two were bound to the protein surface, one being shared between two lysozyme molecules in the renatured crystal. The SDS molecule that penetrated into the hydrophobic core could not be removed even after soaking in SDS-free water for an extended period. Lysozyme has two domains separated by a cleft into which the natural substrate (cell wall polysaccharide) binds. During renaturation the two domains fold separately, trapping the SDS molecule between them.

The structures of complexes lying between the two extremes represented by oxidized globular proteins with a few bound surfactant molecules and reduced globular proteins saturated with surfactant are much more difficult to study, and a variety of models have been proposed and usefully summarized by Ibel et al. [47] as follows:

1. A model based on a "micellar complex" in which the protein assembles the surfactant molecules to form a micelle of definite size.
2. A "rod-like particle" model in which the polypeptide chain forms the core of a rod of 3.6 nm diameter with surfactant bound along its length. The rod is not totally rigid and has flexible regions between short rigid segments.
3. A "pearl necklace" model in which the polypeptide chain forms the string of the necklace and the surfactant molecules form micelle-like clusters along the polypeptide chain, which passes through the micellar clusters in a α-helical conformation. In contrast to the rod-like particle model, this model assumes that the polypeptide chain is flexible.
4. An "α helix–random coil" model in which the surfactant binding enhances the α-helical content of the protein and disrupts the β structure.
5. A "flexible helix" model in which the surfactant molecules form a flexible cylindrical micelle and the polypeptide chain of the protein wraps around it and is stabilized by hydrogen bonding between the surfactant head group and the peptide bond nitrogen atoms.

It is clear from a consideration of the diversity of the proposed models that our understanding of the detailed structure of surfactant–protein complexes is far from adequate despite very many years of study. The pearl necklace model and the flexible helix model could hardly be more different. Neutron scattering has been used in an attempt to get more direct information on the structure of complexes, although here again the results must be fitted to models, and thus the conclusions are not unambiguous. In the case of complexes formed between bovine serum albumin (BSA) and lithium n-dodecylsulphate, the small-angle neutron scattering data were interpreted in terms of the pearl necklace model, in which micelles of 1.8 nm radius (aggregation number 70 ± 20) were distributed along the single polypeptide chain of BSA [48]. The correlations between the micelles are given by the structure factor

$$S(Q) = 1 + N_p \int_0^\alpha 4 \pi r^2 g(r) \frac{\sin Qr}{Qr} \, dr \qquad (29)$$

where N_p is the number density of micelles of radius R and $g(r)$ is the pair correlation function, which is related to $N(r)$, the number of individual scatterers within a sphere of radius r, where

$$g(r) = \frac{1}{4 \pi r^2} \frac{dN}{dr} \qquad (30)$$

$N(r)$ can be related to the fractal dimension D of the protein–surfactant complex by the equation $N(r) = (r/R)^D$. For a freely diffusing micelle the fractal dimension would be 3, but in a complex the distribution of micelles is dictated by the topology of the polypeptide backbone, so that the fractal dimension is less than 3. At 1% by weight of lithium n-dodecylsulphate, D is 2.3 and decreases to 1.76 at 3%, consistent with a transition from a compact state to a more open random coil in which a string of constant-sized micelles are distributed along the hydrophobic patches of a denatured random coil, although the coil will be restrained by the 17 disulphide bonds in the BSA structure.

The pearl necklace model for the BSA–dodecylsulphate complexes is very different from the model used to interpret the neutron diffraction data for the complexes formed between the deuterated bifunctional enzyme N-5′-phosphoribosylanthranilate/indole-3-glycerol-phosphate synthase (PRA-IGP) and SDS [47]. This enzyme from *Escherichia coli* contains a single polypeptide chain of 452 amino acids (molecular weight 49,484) with no disulphide bonds. The deuterated molecule binds 1.26 g SDS per g protein (216 SDS molecules/452 amino acid residues). Neutron scattering was investigated from the whole molecule complex W and two SDS-complexed fragments, produced by gentle hydrolysis with trypsin: a large fragment L containing 289 residues and a small fragment S containing 163 residues. The pair-distance distribution functions of volume elements of the three SDS-complexed structures corresponded to the following; a singular globular structure for the small fragment S with a neutron scattering total dodecyl-chain volume V_c of 26 ± 1 nm^3 and 73 ± 3 C$_{12}$H$_{25}$ chains; a structure with two micelles associated with the C- and N-terminal ends for the large fragment L (V_c $35.6 + 1.5$ nm^3 (42 ± 2 SDS molecules)); and a structure for the whole molecule complex W in which the polypeptide chain of the enzyme is wrapped around three micelles with aggregation numbers 42, 101 (at the C-terminal), and 73 (at the N-terminal). The overall structure for the whole molecule complex is one of a "protein-decorated micelle." In the structure it is assumed that the two interconnecting polypeptide segments, which may bind a small number of SDS molecules, are highly flexible as in the pearl necklace model and the repulsive interaction between the micelles leads to an overall elongated conformation.

There is clearly a considerable difference between the pearl necklace and decorated micelle models, which may in part relate to the differences between the proteins, in particular the fact that BSA has a more restricted conformation because of disulphide linkages. However the most important difference is that in the pearl necklace model the polypeptide chain is believed to pass through micelles of constant size as opposed to around micelles of variable size in the decorated micelle model. However it is interesting that for the decorated micelles formed from the fragments S and L and the whole molecule the numbers of SDS molecules per amino acid residue are surprisingly uniform (0.45 (S), and 0.49 (L), and 0.48 (W)) and very close to the values in the flexible helix model stabilized by hydrogen bonding proposed by Lundahl et al. [110].

A model approximating a decorated micelle is suggested for the complexes formed on interaction of apocytochrome c and its fragments with SDS [11]. Apocytochrome c is the precursor of mitochondrial cytochrome c; it is highly basic and interacts with negatively charged lipids. After synthesis in the cytoplasm it is translocated across the outer mitochondrial membrane by a process coupled to covalent attachment of a heme group catalyzed by cytochrome c hemelyase. During insertion into the membrane it undergoes a random-coil-to-α-helix transition, which has been observed by circular dichroism. From studies in which apocytochrome c has been inserted into phospholipid vesicles, it is found that the amount of α-helix induced is close to 22% for dioleoylphosphatidylserine vesicles; interestingly, a very similar amount of α-helix induction occurs when apocytochrome c interacts with SDS micelles, suggesting that the topologies of the protein in the lipid and SDS micellar environments are similar. Photochemically induced dynamic nuclear polarization ^{1}H NMR spectroscopy of histidine, tryptophan, and tyrosine residues showed enhancement of signals in the presence of SDS micelles, while fluorescence spectroscopy showed quenching of tryptophan and tyrosine by micelles. These results suggested that the aromatic residues are localized in the interface of the SDS micelles, although the possible insertion of the α-helices into the micelle could not be ruled out.

The conformation of salmon calcitonin in SDS micelles has also been shown to be α-helical [112]. Calcitonin is a single polypeptide chain hormone of 32 amino acids with a low content of secondary structure in aqueous solution, but in SDS micelles it has been found to be 50% helical [113]. ^{1}H NMR studies have shown that in SDS micelles residues 6–22 are α-helical and that the 10 residues at the C-terminus form a loop folded back towards the helix. The major coat protein (g VIIIp) of bacteriophage M13 is almost completely α-helical. This 50 amino acid residue protein when in SDS micelles forms two α-helices, a rigid helix from residue 24-45 and a less rigid helix from residue 6-20 [114]. Similarly, the coat protein of phage Pf1 retains its secondary structure in SDS micelles [115], as does recombinant porcine growth hormone in cetyltrimethylammonium chloride [116], and the denatured sulfurtransferase enzyme rhodanese refolds in Triton X-100 or lauryl maltoside mixed micelles with phospholipids [117]. These studies all demonstrate the stability of the α-helical conformation of proteins in the micellar environment. However, despite the relatively extensive investigations that have been carried out on the structure of protein–surfactant complexes, we are still far from predicting what conformation a particular protein will adopt. It is unfortunate that many of the investigations using modern sophisticated physical methods such as neutron diffraction have largely been done without reference to the exact state of binding as represented by the binding isotherm. As the extent of binding changes with surfactant concentration, so will the topology of the complexes, and only when studies are made with careful reference to the binding isotherms will we be able to obtain a clearer picture of the structure of the complexes.

REFERENCES

1. R. R. Fraser, *British Med. J. 2*: 125 (1897).
2. C. Phisalix, *Comp. Rend. Acad. Sci. 125*: 1053 (1897).
3. W. P. Larson, H. O. Halvorson, R. D. Evans, and R. G. Green, The effect of surface tension depressants upon bacterial toxins, in *Third Colloid Symposium Monograph*, 1925, pp. 152–157.
4. M. Baylis, *J. Infect. Dis. 59*: 131 (1936).
5. M. Screenivasaya and N. W. Pirie, *Biochem. J. 32*: 1707 (1938).
6. M. L. Anson, *J. Gen. Physiol 23*: 239 (1939).
7. M. L. Anson, *Science 90*: 256 (1939).
8. M. L. Anson, *Science 90*: 142 (1939).
9. F. W. Putnam, *Adv. Protein Chem. 4*: 79 (1948).
10. J. Steinhardt and J. A. Reynolds, *Multiple Equilibria in Proteins*, Academic Press, New York, 1969, Chap. 6, pp. 234–350.
11. A. Helenius and K. Simons, *Biochim. Biophys. Acta 415*: 29 (1975).
12. O. T. Jones, J. P. Earnest, and M. E. McNamee, Solubilization and reconstitution of membrane proteins, in *Biological Membranes* (J. B. C. Findlay and W. H. Evans, eds.), IRL Press, Oxford, 1987 Chap. 5, pp. 139–177.
13. M. N. Jones, *Biological Interfaces*, Elsevier, Amsterdam, 1975, Chap. 5, pp. 101–134.
14. M. N. Jones, in *Biochemical Thermodynamics*, 2d ed. (M. N. Jones, ed.), Elsevier, Amsterdam, 1979, Chap. 5, pp. 182–240.
15. M. N. Jones, *Chemical Society Reviews 21*: 127 (1992).
16. C. C. Bigelow, *J. Theoret. Biol. 16*: 187 (1967).
17. R. Pitt-Rivers and F. S. A. Impiombato, *Biochem. J. 109*: 825 (1968).
18. J. A. Reynolds and C. Tanford, *Proc. Natl. Acad. Sci. USA 66*: 1002 (1970).
19. E. Dickinson, *An Introduction to Food Colloids*, Oxford Univ. Press, Oxford, 1992, Chap. 2, pp. 47–50.
20. R. D. Swisher, *Surfactant Biodegradation*, Marcel Dekker, New York, 1987.
21. J. V. Moller and M. le Maire, *J. Biol. Chem. 268*: 18659 (1993).

22. K. Weber and M. Osborn, in *The Proteins* (H. Neurath, R. L. Hill, and C.-L. Boeder, eds.), 3d ed., Academic Press, 1979, Vol. 1, pp. 179–223.

23. T. M. Schmilt, *Analysis of Surfactants* (Surfactant Science Series, Vol. 40), Marcel Dekker, New York, 1992.

24. P. Mukerjee and K. J. Mysels, *National Standard Reference Data Series*, National Bureau of Standards (U.S.), No. 36 (1971).

25. C. A. Nelson, *J. Biol. Chem. 246*: 3895 (1971).

26. M. N. Jones, P. Manley, and A. E. Wilkinson, *Biochem. J. 203*: 285 (1982).

27. M. N. Jones, P. Manley, P. J. W. Midgley, and A. E. Wilkinson, *Biopolymers 21*: 1435 (1982).

28. M. N. Jones, A. Finn, A. Moosavi-Movahedi, and B. J. Waller, *Biochim. Biophys. Acta 913*: 395 (1987).

29. F. E. Beyhl, *IRCS Med. Sci. 14*: 417 (1986).

30. M. Y. El-Sayert and M. F. Roberts, *Biochim. Biophys. Acta 831*: 133 (1985).

31. M. N. Jones, *Biochem. J. 151*: 109 (1975).

32. M. N. Jones and P. Manley, *J. Chem. Soc. (Faraday Trans. I) 76*: 654 (1980).

33. M. I. Paz Andrade, M. N. Jones, and H. A. Skinner, *J. Chem. Soc. (Faraday Trans. I) 74*: 2923 (1978).

34. M. N. Jones, A. J. B. MacFarlane, M. I. Paz Andrade, and F. Sarmiento, *J. Chem. Soc. (Faraday Trans. I) 90*: 2511 (1994).

35. M. N. Jones and P. Manley, *Int. J. Biol. Macromol 4*: 201 (1982).

36. M. N. Jones and P. Manley, *J. Chem. Soc. (Faraday Trans. I) 77*: 827 (1981).

37. P. Lundahl, E. Mascher, K. Kameyama, and T. Takagi, *J. Chromatogr. 518*: 111–121 (1990).

38. E. Tipping, M. N. Jones, and H. A. Skinner, *J. Chem. Soc. (Faraday Trans. I) 70*: 1306–1315 (1974).

39. M. N. Jones, H. A. Skinner, and E. Tipping, *Biochem. J. 147*: 229 (1975).

40. M. N. Jones, H. A. Skinner, and E. Tipping, and A. E. Wilkinson, *Biochem. J. 135*: 231 (1972).

41. K. Kale, G. C. Kresheck, and G. Vandekooi, *Biochim. Biophys. Acta 535*: 334 (1978).

42. M. D. Reboiras and M. N. Jones, *Electrophoresis 3*: 317 (1982).

43. M. N. Jones, A. E. Wilkinson, and A. Finn, *Int. J. Biol. Macromolecules 7*: 33 (1985).

44. T. Takagi, J. Miyake, and T. Nashima, *Biochim. Biophys. Acta 626*: 5 (1980).
45. M. N. Jones and P. J. W. Midgley, *Biochem. J. 219*: 875 (1984).
46. S. Makino, R. Maezawa, R. Moriyama, and T. Takagi, *Biochim. Biophys. Acta 874*: 216 (1986).
47. K. Ibel, R. P. May, K. Kirschner, H. Szadkowski, E. Mascher, and P. Lundahl, *Eur. J. Biochem. 190*: 311 (1990).
48. S.-H. Chen and J. Teixeira, *Phys. Rev. Lett 57*: 2583 (1986).
49. A. Yonath, A. Podjamy, B. Honig, A. Sielecki, and W. Traub, *Biochemistry 16*: 1418 (1977).
50. M. N. Jones, *Biochim. Biophys. Acta 491*: 120 (1977).
51. C. Blinkhorn and M. N. Jones, *Biochem. J. 135*: 213 (1973).
52. W. J. Knox and T. O. Parshall, *J. Coll. Int. Sci. 33*: 16 (1970).
53. T. Rades, W. Schütze, and C. C. Müller-Goymann, *Pharmazie 48*: 425 (1993).
54. R. Simha, *J. Phys. Chem. 44*: 25 (1940).
55. P. J. Flory, *The Principles of Polymer Chemistry*, Cornell Univ. Press, 1953, Chap. 14, p. 611.
56. J. Chen and E. Dickinson, *J. Sci. Food Agric. 62*: 283 (1993).
57. E. Dickinson and S. Tanai, *J. Agric. Food Chem. 40*: 179 (1992).
58. A. R. Mackie, P. J. Wilde, D. R. Wilson, and D. C. Clark, *J. Chem. Soc. (Faraday Trans.) 89*: 2755 (1993).
59. D. C. Clark, P. J. Wilde, and D. R. Wilson, *Colloids Surf. 59*: 209 (1991).
60. A. M. Kragh, *Trans. Faraday Soc. 60*: 233 (1964).
61. V. Sovilj, L. Djakovic, and P. Dokic, *J. Colloid Interface Sci. 158*: 483 (1993).
62. A. Samanta and D. K. Chatteroraj, *J. Colloid Interface Sci. 116*: 168 (1987).
63. M. C. Wahlgren and T. Arnebrant, *J. Colloid Interface Sci. 142*: 503 (1991).
64. R. D. Tilton, E. Blomberg, and P. M. Claesson, *Langmuir 9*: 2102 (1993).
65. M. C. Wahlgren and T. Arnebrant, *J. Colloid Interface Sci. 148*: 201 (1992).
66. S. Welin-Klinstrom, A. Askendal, and H. Elwing, *J. Colloid Interface Sci. 158*: 188 (1993).
67. E. G. Cockbain, *Trans. Faraday Soc. 49*: 104 (1953).
68. N. Watanabe, T. Shirakawa, M. Iwahashi, and T. Seimya, *Colloid & Polymer Sci. 266*: 254 (1988).

69. H. Elwing and C.-G. Gölander, *Adv. Colloid Interface Sci. 32*: 317 (1990).
70. M. N. Jones, *J. Colloid Interface Sci. 23*: 36 (1967).
71. H. Nishiyama and H. Maeda, *Biophys. Chem. 44*: 199 (1992).
72. E. Dickinson, *ACS Symp. Ser. 448*: 114 (1991).
73. J.-L. Courthandon, E. Dickinson, and D. G. Dalgleish, *J. Colloid Interface Sci. 145*: 390 (1991).
74. M. N. Jones and A. E. Wilkinson, *Biochem. J. 153*: 713 (1976).
75. M. N. Jones and P. Manley, *J. Chem. Soc. (Faraday Trans. I) 75*: 1736 (1979).
76. M. N. Jones and P. Manley, in *Solution Behaviour of Surfactants* (K. L. Mittal and B. Lindman, eds.), Vol. 2, Plenum Press, 1983, pp. 1403–1415.
77. M. N. Jones, P. Manley, and A. Holt, *Int. J. Biol. Macromol. 6*: 65 (1984).
78. A. Finn, M. N. Jones, P. Manley, M. I. Paz Andrade, and L. Nunez Regueira, *Int. J. Biol. Macromol. 6*: 284 (1984).
79. M. Nogueira, A. Finn, and M. N. Jones, *Int. J. Biol. Macromol. 8*: 270 (1986).
80. A. A. Moosavi-Movahedi, M. N. Jones, and G. Pilcher, *Int. J. Biol. Macromol. 10*: 75 (1988).
81. A. A. Moosavi-Movahedi, M. N. Jones, and G. Pilcher, *Int. J. Biol. Macromol. 11*: 26 (1989).
82. M. I. Paz Andrade, E. A. Rodriguez-Nunez, F. Sarmiento, and M. N. Jones, *Thermochimica Acta 130*: 165 (1988).
83. J. Cordoba, M. D. Reboiras, and M. N. Jones, *Int. J. Biol. Macromol. 10*: 270 (1988).
84. A. A. Moosavi-Movahedi, G. Pilcher, and M. N. Jones, *Thermochimica Acta 146*: 215 (1989).
85. F. Sarmiento, G. Prieto, and M. N. Jones, *J. Chem. Soc. (Faraday Trans.) 88*: 1003 (1992).
86. G. Prieto, J. M. del Rio, M. I. Paz Andrade, F. Sarmiento, and M. N. Jones, *Int. J. Biol. Macromol. 15*: 343 (1993).
87. M. R. Housaindokht, M. N. Jones, J. F. Newall, G. Prieto, and F. Sarmiento, *J. Chem. Soc. (Faraday Trans.) 89*: 1963 (1993).
88. M. R. Housaindokht, A. A. Moosavi-Movahedi, J. Moghadasi, and M. N. Jones, *Int. J. Biol. Macromol 15*: 337 (1993).
89. A. A. Moosavi-Movahedi and S. Ghobadi, *J. Sci. I.R. Iran 3*: 11 (1992).

90. A. A. Moosavi-Movahedi and M. R. Razeghifard, *J. Sci. I.R. Iran* 4: 244 (1990).
91. A. A. Moosavi-Movahedi and M. R. Housaindokht, *J. Sci. I.R. Iran* 4: 253 (1990).
92. A. A. Moosavi-Movahedi, A. Rabbani, M. Goodarzi, and B. Goliaei, *Thermochimica Acta 154*: 209 (1989).
93. A. A. Moosavi-Movahedi and M. R. Housaindokht, *Int. J. Biol. Macromol. 13*: 50 (1991).
94. A. A. Moosavi-Movahedi and M. R. Razeghaifard, *J. Sci. I.R. Iran* 2: 15 (1991).
95. A. V. Hill, *Biochem. J. 1*: 471 (1914).
96. G. Scatchard, *Ann. N.Y. Acad. Sci. 51*: 660 (1949).
97. M. N. Jones and A. Brass, in *Food Polymers and Colloids* (E. Dickinson, ed.), Royal Society of Chemistry Special Publication No. 82, 1990, pp. 65–80.
98. G. Schwarz, *Biophys. Struct. Mech. 2*: 1 (1976).
99. I. M. Klotz and D. L. Hunston, *J. Biol. Chem. 259*: 10060 (1984).
100. I. M. Klotz, *Science 217*: 1247 (1982).
101. H. A. Feldman, *J. Biol. Chem. 258*: 12865 (1983).
102. J. Wyman, *J. Mol. Biol. 11*: 631 (1965).
103. M. I. Paz Andrade, M. I. Boitard, M. A. Saghal, P. Manley, M. N. Jones, and H. A. Skinner, *J. Chem. Soc. (Faraday Trans I)* 77: 2939 (1981).
104. W. Pfeil, in *Biochemical Thermodynamics* (M. N. Jones, ed.), Elsevier, Amsterdam, 2d ed., 1979, Chap. 3.
105. M. N. Jones, E. M. Gilhooly, and A. R. Nicholas, *J. Chem. Soc. (Faraday Trans.) 88*: 2733 (1992).
106. B. L. Roberts, J. F. Scamehorn, and J. H. Harwell, in *Phenomena in Mixed Surfactant Systems* (J. F. Scamehorn, ed.), ACS Symposium Series No. 311, American Chemical Society, Washington D.C., 1986, pp. 200–215.
107. J. H. Clint, *J. Chem. Soc. (Faraday Trans. I) 71*: 1327 (1975).
108. M. N. Jones, P. Manley, and P. J. W. Midgley, *J. Coll. Int. Sci. 82*: 257 (1980).
109. K. J. Mysels, *J. Colloid Sci. 10*: 507 (1955).
110. P. Lundahl, E. Greijer, M. Sandberg, S. Cardell, and K.-O. Eriksson, *Biochim. Biophys. Acta 873*: 20 (1986).
111. M. M. E. Snel, R. Kaptein, and B. de Kruijft, *Biochemistry 30*: 3387 (1991).

112. A. Motta, A. Pastore, N. A. Goud, and M. A. Castiglione Morelli, *Biochemistry 30*: 10444 (1991).
113. R. M. Epand, R. F. Epand, and R. C. Orlowski, *Int. J. Pept. Protein Res. 25*: 105 (1985).
114. F. J. M. van de Ven, J. W. M. van Os, J. M. A. Aelen, S. S. Wymenga, M. L. Remerowski, R. N. H. Konings, and C. W. Hilbers, *Biochemistry 32*: 8322 (1993).
115. R. A. Schiksnis, M. J. Bogusky, and S. J. Opella, *J. Mol. Biol. 200*: 741 (1988).
116. M. Cardamone, N. K. Puri, W. H. Sawyer, R. J. Capon, and M. R. Brandon, *Biochim. Biophys. Acta 1206*: 71 (1994).
117. G. Zardeneta and P. M. Horowitz, *Anal. Biochem. 218*: 392 (1994).

9
Factors Affecting Applications of Native and Modified Proteins in Food Products

M. E. Mangino and W. James Harper

The Ohio State University, Columbus, Ohio

I. EMULSIONS

A. Introduction

The basic forces and types of interactions involved in emulsion formation were discussed in Chap. 1. To summarize briefly, perturbations in the structure of water, when it is forced into contact with nonpolar substances, lead to a condition of high free energy. Energy is minimized when the area of contact is decreased. The condition of lowest free energy for two liquids of greatly different polarities, such as oil and water, occurs when they separate into different phases. Emulsion formation in foods depends on the use of amphiphilic molecules for which a portion of the molecule is soluble in water while the other portion is soluble in nonpolar solvents. These agents lower the energy of the system and result in a metastable dispersion of one phase in a continuous phase of the other. The phases will remain dispersed as long as the distance of separation of the dispersed phase is great enough to prevent coalescence. Given enough time, the system will reach the state of lowest free energy when the phases completely separate. The task of the processor of emulsified products is to ensure that the time required for the inevitable phase separation is greater than the shelf life of the product.

Factors that are important to the quality of food emulsions include the type and amount of fat in the product, the types and amounts of small emulsifier molecules present in the product, the temperature and pressure of homogenization, the type and amount of protein present, the final composition of the product, and the conditions of storage. These variables are important during emulsion formation as well as in subsequent storage of the emulsion. While many of the considerations of emulsion formation and storage are similar, the differences are great enough to justify discussing them separately.

B. Emulsion Formation

The formation of an emulsion requires the input of work greater than the increase in energy that results from the increased area of surface contact. According to Tornberg and Hermansson [1] the method of emulsion formation is the most important factor in determining the nature of the emulsion. Tornberg et al. [2] have reviewed methods of emulsion formation and the merits of each procedure. Walstra [3] has comprehensively discussed the mechanisms and forces required to create an emulsion. In all of the methods used to form emulsions, the amount of interfacial area created is a function of the amount of work done on the system. Formation of small droplets requires a greater input of work than does that of large ones. The system attempts to lower its energy through the coalescence of fat globules. The rate of droplet collision and the presence or absence of an energy barrier will determine the rate of fat globule coalescence. The energy barrier to coalescence for uncoated fat globules is small enough to be ignored.

The size distribution of emulsions is controlled by the rate of diffusion of functional molecules to the interface [4]. Emulsifier molecules diffuse to the interface while coalescence is occurring [4]. If the coating of fat globules were instantaneous, the emulsion would have a particle size distribution identical to that at the moment of emulsification. In actuality, emulsifier molecules require a finite time to reach the interface and be adsorbed. The rate of droplet coating is determined by (1) the rate of fat droplet coalescence, (2) the rate at which emulsifiers reach the interface, and (3) their rate of absorption. The rate of diffusion of small surface-active molecules is much greater than that of protein emulsifiers. Additionally, the amount of reorientation required to interact at the surface is less for small molecules. In general, the use of small emulsifier molecules results in a relatively narrow particle size distribution and in the formation of emulsions with relatively small fat globules [5].

Factors that affect the rate of protein diffusion will affect their suitability as emulsifiers. In order for a protein to be able to diffuse it must be either soluble or easily dispersible. The importance of protein solubility to emulsion formation has been reviewed by Halling [6]. Protein solubility has been said to be the primary factor governing protein functionality [7,8]. However, for many applications, including emulsification, it has been demonstrated that over reasonable ranges of solubility (35 to 95%), protein solubility is often not the primary factor in the determination of protein functionality [9,10]. Mangino [11] has reviewed available data and has concluded that complete protein solubility is not required for adsorption onto the fat globules surface. Proteins that are not totally soluble may still function effectively in emulsion formation. However, a protein must be able to interact with other molecules and thus must be able to be dispersed in the continuous phase. Some proteins with limited solubility in water may show improved dispersibility when their hydrophobic groups are brought into contact with fat.

The final pH of the emulsion will determine to a large extent which proteins will be functionally effective as emulsifiers. Halling [6] suggested that the effects of pH on the emulsifying behavior of proteins could be explained by one or more of the following determining factors:

1. The isoelectric point is the pH of minimal protein solubility, and for many proteins the reduced solubility may affect emulsification functionality.
2. The cohesiveness of protein films tends to be maximal near the isoelectric point. Cohesive films tend to be more stable.
3. Charge repulsion of emulsion droplets is minimal near the isoelectric point, resulting in decreased stability.
4. Near the isoelectric point, repulsion of charged segments of pro-

tein chains is minimized, allowing for the adsorption of compact protein structures.

Proteins that are not soluble at their isoelectric point cannot readily participate in emulsion formation. Thus proteins such as casein and soy are not suitable for emulsified products with a pH near 4.6. For proteins that are soluble at their isoelectric points, this may be the pH of maximal protein adsorption [12]. Near the isoelectric pH, electrostatic repulsion is minimized, allowing hydrophobic residues to stabilize a more compact tertiary structure. There are data to support that for some proteins, absorption is increased at pH values on either side of the isoelectric point. If enough work can be expended to obtain protein adsorption, charge repulsion due to the residual charge increases emulsion stability.

The rate and extent of protein unfolding at an interface is related to its charge distribution, the number and distribution of hydrophobic groups present, and its flexibility. Caseins are extremely flexible molecules with little secondary or tertiary structure [13] and are excellent emulsifiers. In addition, their charged groups tend to be localized, and they possess large regions that are rich in hydrophobic groups. Hydrophobic amino acids are necessary to interact with the lipid phase. Hydrophobic residues are most reactive if there are no charged groups in close proximity. MacRitchie [14] suggested that a minimum of from six to eight adjacent hydrophobic amino acids is required for insertion of a hydrophobic region into a nonpolar environment. Close proximity of charged amino acids to hydrophobic ones will interfere with protein adsorption. Therefore localization of an adequate number of hydrophobic residues allows their insertion into the lipid phase. This process has a low energy of activation and proceeds spontaneously [15].

Molecules that contain significant amounts of secondary and tertiary structure, such as β-lactoglobulin, require greater amounts of time to achieve comparable surface coverage. Studies of the rate of increase of surface pressure with time for various proteins have demonstrated that flexible molecules show little lag between their initial absorption and changes in surface pressure. Proteins with considerable structure require much greater times from their initial absorption to maximal surface pressure increase [16]. This lag has been suggested to result from the unfolding of structured areas of the proteins already adsorbed at the interface [17]. It is likely that β-lactoglobulin is adsorbed as a compact globular molecule possessing considerable tertiary structure. Given time, more hydrophobic groups within the molecule are able to come into contact with the interface and become adsorbed.

Proteins like soy that contain rigid quaternary structures are even more difficult to unfold. Soy shows an even greater lag than β-lactoglobulin in surface pressure increase following absorption [18,19]. To obtain rapid surface coverage of fat globules with soy proteins requires that emulsification occur under conditions that disrupt quaternary structure. With the application of sufficient work, soy proteins form better emulsions at pH values far above their isoelectic points. High energy input is required to overcome the increased charge on the proteins and the increased solution viscosity due to protein unfolding. High viscosity of the continuous phase decreases the rate of diffusion to the interface.

From the above discussion it is possible to generate some general guidelines regarding the selection of proteins for the formation of emulsions. The most efficient formation will occur with proteins that are flexible at pH values near their isoelectric points. The most flexible of the commonly utilized food proteins are the caseins. Unfortunately, they are not soluble in the isoelectric range and their use is limited to pH values approximately one unit above or below their isoelectric points. Emulsions formed with casein will tend to have smaller average fat globule diameters than those made with less flexible proteins due to more efficient absorption of casein to the interface [20]. From a theoretical standpoint, this should result in higher quality emulsions. However, as will be discussed later, the stability of emulsions in food systems depends on more factors than the average fat globule diameter. In certain applications, the properties of casein that allow it to form such uniform emulsions are not those that in general give greatest emulsion stability.

Emulsion formation is usually optimal at temperatures around 60°C, coinciding with the temperature at which the interactions of hydrophobic groups are optimal [21]. Thus, at this temperature, interactions between emulsifiers and lipids as well as the hydrophobic groups of proteins and lipids will be maximized and will favor the conditions that promote stable emulsions. Another effect of increased temperature is to decrease the viscosity of the continuous phase during processing. Decreased viscosity will enhance the diffusion of molecules to the interface and increase the efficiency of emulsification. Temperatures much above 60°C provide no additional benefit and may be detrimental. While the viscosity will continue to decrease with further increases in temperature, the weakening of hydrophobic associations between emulsifier molecules and the lipid phase negate such gains. Changes in the structure of water and the nature of its hydrogen bonding at higher temperatures are the likely cause for the decrease in the strength of hydrophobic associations at elevated temperatures [21]. In practice, the temperature rise due to mechanical friction may have to be considered when the optimal temperature of homogenization is determined. At very high pressures (430 bars) an almost instantaneous temperature rise of approximately 20°C may be experienced. If the material to be processed contains heat-sensitive proteins, these could be irreversibly denatured with a resulting loss of functionality.

C. Emulsion Stability

Emulsion failure usually results from coalescence, flocculation, gravitational creaming, or Ostwald ripening. Ostwald ripening results from the diffusional transport of material from smaller droplets into larger ones. In protein stabilized emulsions this is probably not of great significance [11]. Flocculation refers to the aggregation of individual fat globules into large clusters. Often, this results from the interaction of divalent cations with susceptible proteins. The addition of excess calcium to a casein stabilized emulsion, for example, will lead to flocculation. Generally, flocculation is less serious than coalescence, but it may present a significant problem when the attractive forces between the particles are strong enough to prevent easy redispersion of the system.

Coalescence results when fat globules collide with enough energy to fuse. The forces that are important to the coalescence of fat droplets can be approximated by use of the DLVO theory [22]. This theory suggests that in order to have a stable suspension, the attractive forces must be offset by a set of repulsive ones. The main attractive forces are transient dipole-induced dipole interactions. These are relatively weak forces that become significant when the distance of separation between fat globules is small. Their strength is proportional to the distance of separation of the lipid droplets to the power of minus 6. Random motion in the dispersed phase can cause individual lipid droplets to come fairly close unless a repulsive force is at work.

The DLVO theory predicts that this repulsive force will be roughly equivalent to the residual charge on the surface of the lipid droplets. It has been calculated, however, that the charge repulsion required to overcome the attractive van der Waals forces for food emulsions is rarely present at the surface of the lipid droplets [23]. It has been suggested by Friberg et al. [24] that the formation of liquid crystalline regions of emulsifier molecules may add to stability. The conditions required for the formation of liquid crystalline structures depend upon the ratios of oil, emulsifier, and water, the temperature of the system, and the nature of the emulsifier molecules. Concentrations of emulsifier in the range of 3 to 6% of the total lipid concentration have been reported to be required for the formation of liquid crystals. The formation of liquid crystals increases the distance of separation of hydrophobic groups and thus decreases the strength of van der Waals attractive forces. Problems with coalescence are generally encountered in relatively high fat products such as salad dressings. In these types of products, the formation of liquid crystals may add significant emulsion stability.

Creaming refers to the concentration of lipid in the upper portion of a fluid emulsion. Differences in density between the dispersed and continuous phases are responsible for creaming. The rate of creaming can be approximated by the Stokes law,

$$v = \frac{2r^2 g \Delta p}{9\mu}$$

where v equals the velocity of the fat globule, r is the radius of the fat globule, g is the force of gravity, Δp is the density difference between the two phases, and μ is the viscosity of the continuous phase. For emulsions with an average fat globule diameter of less than 0.1 micron, the rate of creaming is slow enough to be ignored. Few food emulsions would be expected to have average fat globule diameters in this range. Fluid products characterized by an average fat globule diameter of 1 micron or greater can be expected to experience significant rates of creaming.

Decreasing the average fat globule diameter or the density differences between the two phases may be expected to inhibit creaming. Increasing the viscosity of the continuous phase would have a similar effect. Both the type and the amount of emulsifier present as well as the method of emulsion formation will affect the average fat globule diameter [4]. Most food emulsions are represented by fat globule diameters of less than one micron. It is doubtful that for most food products much improvement in controlling diameter size is possible.

Density differences between the two phases are difficult to control. The range of fat globule diameters obtained during emulsion formation is too wide to allow the processor predictably to adsorb controlled amounts of protein and negate density differences. In a few applications, especially with essential flavorings, it has been possible to increase the density of the lipid phase with the addition of substances such as brominated vegetable oil. This approach is not suitable for most emulsified products.

Some products have sufficiently viscous continuous phases to inhibit gravitational creaming. Examples would include salad dressing and mayonnaise. For products that are meant to be consumed as beverages, such as milk, increased continuous phase viscosities are not possible. Even though manipulation of continuous phase viscosity has only limited applications, rheological considerations are still important. Typical emulsions are pseudoplastic with viscosities that decrease rapidly with shear. Formation of gellike networks having sufficiently low yield stresses can increase stability towards creaming. Walstra [3] calculated that yield stresses in excess of 0.1 Pa would be adequate to prevent creaming. Yield stresses of 10 Pa or less would allow flow upon pouring.

Caseins are often used in emulsions that require considerable thermal stability. The resistance of these proteins to denaturation allows processing under conditions that would cause gelation if whey, egg, or soy proteins were used. Their resistance to thermal denaturation prevents their participation in network formation. Thus while they impart excellent heat stability, emulsions formed with casein are susceptible to creaming. Carrageenan is often added to casein stabilized emulsions to enhance their stability towards creaming. It has been suggested that the ability of carrageenan to interact with specific caseins and form a gellike network is responsible for their increased stability [25,26]. In applications where both heat stability and longterm stability towards creaming are essential, combinations of casein and other proteins are often effective. Fligner et al. [27] reported that as the ratio of whey proteins to casein in model infant formulas increased, the amount of creaming also decreased. The increased stability imparted by whey proteins was attributed to the formation of weak protein gels. Complete replacement of casein with whey was not possible in this system, as a certain level of casein was required to prevent total gelation of the product during thermal processing. In a recent paper, Sharma and Dalgleish [28] reported that for emulsions formed with milk at 40°C there was no absorption of whey proteins on the fat globule surface. Significant whey protein absorption occurred only upon subsequent heating of the emulsions. It would be interesting to know the extent of whey protein absorption that occurs when homogenization is performed at 60°C. The ability selectively to control the extent of whey protein absorption to the fat globule surface might have important implications as to emulsion stability as discussed below.

Virtually all food emulsions are formed with a combination of small emulsifier molecules and proteins, making the situation highly complex. In mixed systems a competition for surface absorption will occur. Small molecules can diffuse faster than large ones and can be expected to cover a greater percentage of the surface when mixed emulsions are formed. Dickinson and coworkers [29,30,31,32] have studied the effects of various emulsifier molecules on the fat globule size distribution, surface rheology, and protein load of a number of milk protein stabilized emulsions. The results depend upon the lipid source, protein, and emulsifier molecule studied. For example, more β-lactoglobulin was absorbed to pure *n*-tetradecane than to soybean oil [30]. Some general observations regarding the ratio of emulsifier to protein and the nature of the emulsions formed could be made. As the ratio of emulsifier to protein increases to about 16, the protein load on the fat globules decreases. This is accompanied by a decrease in the surface viscosity of the interfacial film and a slight decrease to no change in the average fat globule diameter. These changes can have significant effects on the long term stability of the emulsions formed. In addition to the amount of emulsifier present, the type of emulsifier has been shown to affect the quantity of whey protein adsorbed at lipid interfaces [33].

The effect of emulsifier content on the long-term stability of model infant formula emulsions was reported by Fligner et al. [34]. Emulsion stability was found to increase with added lecithin in a system containing casein and whey proteins up to a lecithin concentration of 0.5% of the total fat. Further addition of lecithin caused a decrease in the stability of the emulsion under these conditions. The amount of protein absorbed to the fat globule surface showed a similar maximum at 0.5% lecithin. The displacement of protein from the interface had a negative effect upon the yield value of the system and made the fat globules more susceptible to creaming. Thus while relatively high levels of emulsifier can increase the stability of emulsions that contain high levels of fat and exhibit relatively high viscosities through liquid crystal formation, the displacement of lipid from the fat globule surface may decrease the stability of emulsions towards creaming.

D. Protein Modifications to Enhance Emulsification

There are numerous reports in the literature describing the effects of protein modification on their emulsifying properties. Stainsby [17] has classified these modifications as being chemical or enzymatic in nature. Chemical modifications are often performed to increase the charge and thereby the solubility of proteins. However, modifications are often very drastic, and it is unlikely that such modified proteins will receive regulatory approval for use in food in the near future. The work of Ponnampalam et al. [35] is a representative example of chemical modifications. In this study, oat proteins were either acylated or succinylated to modify approximately 35 and 75% of the amino groups of lysine. In addition, approximately 12 to 50% of the cysteine sulfhydryl groups were also modified. The area of fat emulsified per unit of protein increased by from 25 to 50% for the modified proteins. The values obtained for the oat proteins that were succinylated at 76% substitution approximated those obtained for unmodified whey protein concentrates. The stability of the emulsions formed by the modified proteins was increased to approximately the same extent, and the stability of all of the emulsions produced with modified proteins was greater than the stability of emulsions formed with native whey protein concentrates. All of the modified proteins were more soluble in the isoelectric region than were the unmodified proteins.

Recently, there has been interest in modification of proteins by the incorporation of carbohydrate groups. So far this has often been accomplished by chemical means; however, advances in enzymatic modification may be expected in the future. Such modification would likely be more acceptable to regulatory agencies. As an example, Waniska and Kinsella [12] compared the functionality of native and glycosylated β-lactoglobulin in a number of systems. Their research focused primarily on the foam forming properties and on emulsion formation. In general, the modified proteins had better emulsifying properties than did the native protein. The greatest improvements in functionality were found near the isoelectric point. The stability of the emulsions formed with the modified proteins was slightly improved.

Surface properties of proteins have also been altered by covalent attachment of hydrophobic groups. Of particular interest is a recent report by Haque [37]. Whey proteins and peptides were modified by noncovalently attaching lipids of controlled chain length by a proprietary procedure. These modifications have been reported to substantially enhance the functionality of these molecules at interfaces. Low-fat ice creams and whipped toppings that incorporate these molecules have been reported to have enhanced consumer acceptability. The regulatory fate and the application of these modified proteins and peptides remain uncertain.

Jimenez-Flores and Richardson (36) have discussed a strategy for the rational modification of proteins using a genetic engineering approach. They have suggested that the rapid advances that have been made in this area justify future modifications in protein composition. If work with model systems can identify a carbohydrate-based modification that greatly enhances protein functionality, then the production of the modified protein through genetic engineering may be economically feasible.

II. FOAMING

A. Foam Formation

The formation of a foam is analogous to the mechanism of emulsion formation. The dispersion of either air or oil in an aqueous system involves similar forces. Many of the properties of proteins that contribute to their functionality on emulsions are identical to those that contribute to functionality in foams. Because of these similarities, this discussion will be limited to those areas that are either different or of extreme importance in foaming. In the formation of a foam, water molecules surround air droplets. Major differences between emulsions and foams relate to the relative volumes and densities of the dispersed phases. In foams, the volume of the dispersed phase is much greater than that of the continuous phase. In emulsions, the dispersed phase is usually less than or about equal to the continuous phase [38]. When water molecules are forced into contact with relatively nonpolar air, they tend to become more ordered. This results in a high surface tension and a high surface energy. The action of emulsifiers in foams is analogous to their action in emulsions [39]. To create a stable foam, it is necessary that molecules be present that can lower the interfacial energy and provide a kinetic barrier to air cell coalescence. Proteins that stabilize foams require many of the same properties as are required to form an emulsion.

A newly created foam will contain a number of relatively small air cells surrounded by an interfacial layer of amphiphilic molecules and areas of bulk water. Like an emulsion, a foam is thermodynamically unstable. A number of forces will act to stabilize and destabilize the foam. These forces have been summarized by Kinsella and Whitehead [40] and are listed in Table 1. Gravity will cause the bulk water to drain as the foam ages, causing air cells to approach each

Table 1 Factors Affecting the Stability of Foams

Stabilizing factors	Destabilizing factors
Increased surface viscosity	Gravitational drainage
Increased film thickness	Capillary pressure drainage
Gibbs–Marangoni effect	Mechanical shock
Net surface charge	Film permeability
Residual tertiary structure	Surface active molecules

Source: Modified from Kinsella and Whitehead, 1989, Ref. 40.

other. As the distance of separation of air cells decreases, the phenomenon known as capillary drainage works to thin further the walls separating them. Mechanical disturbances can cause air cells to collide and rupture, resulting in the formation of a smaller number of larger cells. Proteins that bind water tightly and are rigid increase surface viscosity, which tends to reduce the rate of water drainage [41]. If the viscosity reaches a critical value, all flow will stop and the film will be stabilized. Thus proteins that are rigid and exhibit a high surface viscosity tend to form more stable foams [42]. Residual charge is important to the stability of protein foams. The close approach of air cells upon film drainage can lead to attraction due to Van der Waals forces or to charge repulsion if the surface contains areas of like charges. Residual charge on the emulsifier molecules can be very important to film stability, but excessive charge may lead to repulsion during the formation of the film and to incomplete coverage by the protein molecules [43].

The Gibbs–Marangoni effect results from the increased surface tension that occurs in areas of water drainage. The drainage of water also removes some of the surface active interfacial material. The resulting increase in surface tension, and the increase in concentration of emulsifier molecules in areas adjacent to the location of water drainage, causes a diffusion of emulsifier back to this location. The emulsifier molecules carry water with them, and the thickness of the film is restored. This self "healing" of films is very important to their stability (Mangino [4]).

The properties of proteins that promote emulsion formation are also important in the creation of foams. Poole and Fry [44] suggested that the ideal protein for foaming would posses high surface hydrophobicity, high solubility, and a low net charge at the pH of the food product. To exhibit functional performance, the protein must reach the air/water interface. Rapid diffusion and unfolding at the interface is required to lower the interfacial tension between the air and the water phase [45]. Factors that increase the rate of protein diffusion have also been reported to enhance protein foaming [46,16].

As with emulsion formation, the maximal absorption of proteins has been reported to occur near the isoelectric point [6,18,47]. The effects of pH on the properties of proteins at emulsion surfaces, noted previously, probably also apply to the behavior of proteins in films. While foams formed near the isoelectric point of proteins tend to be more stable than those formed at other pH values, it is often possible to obtain higher overruns away from the isoelectric point [48]. The largest effect of pH on foam stability can probably be explained by the more viscoelastic nature of the films formed in this pH region [49].

B. Modification of Proteins to Enhance Foaming

Heating of proteins in both the presence and the absence of water has been reported to increase their foaming properties. Devilbliss et al. [50] reported that heating caused an increase in the foam stability of whey protein concentrates. Kinsella [8] suggested that the effect of mild heating may be to cause a partial unfolding of the protein molecules, which makes the intermolecular interactions that are necessary for stable foam formation easier to form. It has been reported that heating of whey protein concentrates at temperatures of from 50 to 60°C caused a reversible improvement in their foaming properties [51]. There are probably two mechanisms that can explain these observations. Improvement in the surface functionality of egg white, β-lactoglobulin, and α-lactalbumin have been noted for these proteins following heating at 80°C for several days at low moisture contents [52,53]. The authors hypothesized that heating caused the formation of a molten globulelike structure. Proteins in this state would be partially unfolded and more flexible than proteins in the native state. Presumably, absorption of partially unfolded proteins at the air/water interface was more efficient. The authors suggested that such treatments might be useful to improve the functionality of WPC. The improvement noted by the authors occurred with either isolated whey proteins or egg proteins that had their carbohydrate content reduced by enzymatic treatment prior to drying. Heating of WPC without these precautions would likely result in considerable nonenzymatic browning that would tend to decrease protein functionality.

Cooling heated whey protein suspensions before whipping tends to reverse the improvement noted for overrun. Richert et al. [51] speculated that heating might disrupt protein–lipoprotein complexes. Haggett [54] also demonstrated that storage at 4°C caused a decrease in the foaming properties of whey protein concentrates that was completely reversible by mild heat treatment. It is likely that heat treatment of protein solutions containing residual lipids causes the formation of hydrophobically stabilized complexes between the lipid and the protein. The formation of these complexes would effectively reduce the free lipid content of the suspension and increase overrun. Cooling to 4°C would weaken the complex and allow free lipid to interfere with foam formation upon whipping.

Phillips et al. [55] have reported that the functionality of whey protein concentrates and isolates in foaming applications can be greatly enhanced by microfiltration. The major effect of microfiltration was shown to be the removal of lipoprotein complexes found in these systems. These complexes have been reported to be derived from fragments of the original milk fat globule membrane that surrounds native lipid droplets. Some of these fragments are extremely detrimental to the foaming of whey proteins [56,57]. The processes described above are not major modifications of proteins. Because they are typical of the processes that foods normally receive, there should be no regulatory problems associated with their implementation. WPC products, treated by microfiltration to reduce lipid content and increase performance, are already commercially available.

Cleavage of disulfide bonds has been reported to improve the functional performance of proteins in foaming applications [58,59,60]. Cleavage of disulfide bonds in soy proteins has been speculated to improve foaming by reducing the number of intersubunit disulfide bonds [58]. In the case of whey proteins, the reduction of the number of intramolecular bonds results in a more flexible molecule with a greater hydrodynamic size [49]. The types of reagents that cleave disulfide bonds are toxic and impart strong off flavors. Research is needed to demonstrate that the cleavage of one or a few disulfide bonds significantly enhances foaming. If so, it may be possible to engineer proteins genetically with appropriate amino acid substitutions.

The importance of protein–protein interactions to the stability of emulsions has been noted. Phillips et al. [61] have suggested that the excellent stability of egg white foams may be due, in part, to an interaction between lysozyme and other egg white proteins. At the pH of egg white, lysozyme has a net positive charge and is capable of forming electrostatic interactions with negatively charged proteins. Poole et al. [62] added basic proteins (clupine or lysozyme) to negatively charged proteins in a foaming system. They reported dramatic increases in overrun and foam stability. Phillips et al. [61] found that the addition of lysozyme to either β-lactoglobulin or whey protein isolate doubled the time before significant drainage occurred. The heat stability of the mixtures was also increased. German and Phillips [38] have suggested that the use of mixtures of positively and negatively charged proteins has the potential to increase the commercial utilization of proteins in foaming applications. It might be possible, for example, to increase the amide content of proteins to give them enough positive charges to find functionality in such systems.

III. GELATION

A. Introduction

While gelation is not strictly a surface phenomenon, it is very important to the function of a number of foods in both emulsions and foams. Often the stability of a foam or an emulsion can be directly related to the ability of proteins to form stable gels under appropriate conditions. Protein gels may vary greatly in their appearance and rheological properties. According to Ziegler and Foegeding [63] a gel is a continuous network of macroscopic dimensions immersed in a liquid and exhibiting no steady state flow. This definition is relevant to the stability of foams and emulsions. Other descriptions of gels include a requirement that they be viscoelastic, providing an objective measure for their rheological characteristics. Proteins can be induced to form gels by appropriate changes in pH, through the action of certain enzymes, by the addition of salts, or by the application of heat [43]. Most of this paper will deal with thermally induced gel formation.

Gelation is generally a two-stage process involving an initial unfolding of a protein molecule followed by subsequent aggregation. It has been reported to involve a pre-gel stage, a gel point, and a post-gel stage [64]. In the formation of a pre-gel, protein monomers form soluble protein aggregates. The formation of a highly ordered gel requires that the rate of protein aggregation be slower than the rate of unfolding [65]. Heat treatment of proteins weakens the bonds that maintain their secondary and tertiary structures. When thermal denaturation occurs, proteins unfold and water binding increases. When conditions favor the formation of protein–protein interactions, a three-dimensional network capable of entraining water molecules and a gel is likely to form [39]. If the network is too weak, a viscous solution will occur. If the interactions become too extensive, a collapse of the network and subsequent expulsion of water may result [39]. Thus the attractive and repulsive forces necessary to form a network and the repulsive forces necessary to prevent its collapse must be balanced for optimal for gel formation.

B. Factors Important to Gel Formation

Factors reported to be important in the gelation of proteins include pH, salt concentration, calcium concentration, protein hydrophobicity, and free sulfhydryl concentration [39].

Calcium is capable of forming cross-links with a number of food proteins. Some proteins, such as casein and soy proteins, can be precipitated in the presence of appropriate amounts of calcium. Undenatured whey proteins are not calcium sensitive, but the concentration of calcium can greatly influence their gelation properties. Some properties of whey protein gels have been related to ionic concentration. Schmidt [66] reported that the firmest whey protein gels were formed at 11 mM CaCl2 or 200 mM NaCl. At neutral pH values, gel firmness was more sensitive to calcium than to disulfide interchange reactions [67,68]. Lupano et al. [69] hypothesized that low levels of ionic calcium enhanced protein–protein interactions and made a positive contribution to the formation and strength of the gel network. At higher concentrations, calcium caused excessive protein aggregation and precipitation. The presence of aggregated protein has a negative effect on the cohesiveness and firmness of the gels. In the recent studies of Barbut and Foegeding [70], the addition of 20 mM CaCl$_2$ to preheated WPI induced the gelation of whey proteins at room temperature. The ability to form gels at reduced temperatures may have considerable applications in foods.

The effect of sulfhydryl content on the strength and texture of WPC gels was described by Schmidt and coworkers [66,71]. As observed with calcium, there is an optimal sulfhydryl concentration for gel strength. The texture of gels can be changed with the addition of thiol reagents [72]. Langley and Green [73] demonstrated an increase in compressive strength, elastic modulus, and impact strength of whey protein solutions increased with β-lactoglobulin content. Kim et al. [74] found that a correlation existed between the sulfhydryl content of WPC and the β-lactoglobulin content. Dunkerley and Hayes [75] and Schmidt [66] reported that a pH dependence existed between the appearance and the strength of the gels formed from whey protein concentrates. Gels formed at low pH values tended to be soft and opaque and were described as coagula. At higher pH values the gels became more elastic and transparent and had greater gel strengths. A strong pH dependence of the effects of sulfhydryl groups on the strength of WPC gels was reported by Mangino et al. [76]. They confirmed the observation of Kohnhorst and Mangino [77] that at pH values below about 7, there was little effect of sulfhydryl content on gel strength. At higher pH values the effects of sulfhydryl groups became significant.

The relationship of protein hydrophobicity to the gelation characteristics of proteins has been reported by a number of workers [57,77,78,79,80]. There appears to be an optimal level of hydrophobicity beyond which gel strength is weakened. With commercially available whey protein concentrates, the optimal level, as measured by alkane binding, was not reached, and there was a direct positive correlation between gel strength and protein hydrophobicity [77]. Hydrophobic associations provide junction zones for protein chain cross-linking and would be expected to increase gel strength. However, too many such sites could lead to excessive protein aggregation. Studies with commercially available WPC have demonstrated that increases in protein hydrophobicity will increase the strength of the gels formed. Only in a model system of β-lactoglobulin and added milk fat globule membrane [57] has the level of hydrophobic groups been high enough to lower gel strength. It is difficult to alter reproducibly the hydrophobicity of protein solutions. Mangino et al. [81] suggest that a positive effect may be obtained by decreasing the lipid content of the WPC. This would free hydrophobic sites on the proteins and make them available to interact during gel formation. This speculation is consistent with the observation of Mulvihill and Kinsella [82] that lipids impair gelation.

IV. APPLICATIONS OF PROTEINS IN FOODS

A. Introduction

Food proteins are important in determining the characteristics of many food products. Frequently the protein used influences more than one characteristic of the food. The protein selected will vary as a function of the protein, the formulation of the food, and the processing of the product. The most common proteins used as food ingredients include egg proteins [83,84], soy proteins [85,86], milk proteins [87,88,89], wheat gluten [90], and fish proteins [91]. Other proteins have been used to a lesser degree and include rapeseed protein, sunflower protein, pea protein, cottonseed protein, peanut protein, and blood plasma.

Although the functional properties of proteins in simple aqueous systems may be complex, their application in complete food systems is an even more complex issue, and understanding is still incomplete and inadequate. Up until the present, successful application of proteins as food ingredients was more an art than a science. However, there have been solid gains in the understanding of the functional properties of proteins for many applications, which have been translated into a variety of new products with considerable consumer appeal. For example, low-fat and no-fat foods have been formulated on the basis of knowledge of the application of protein technology. Product development efforts on fat-reduced desserts, bakery products, and restructured meats have created a field of endeavor that calls for rigorous research on food protein technology. Results from testing protein functionality in aqueous solutions do not accurately predict the functionality of proteins in foods [92]. Model food systems [89,92,93] have been useful in providing a better understanding of the factors that influence protein behavior, but they do not often translate to commercial practice. This is primarily because they are multifunctional ingredients, and so their performance is modified by component interactions and by the processing of the food [91]. Newer statistical methods, such as neural networks [94], simplex-centroid [95] techniques, and random-centroid [96] techniques may be useful to bring us closer to an understanding of the relationship of protein structure to function in real food products. These techniques may also be useful in understanding complex interactions and for formulation of protein foods in which up to 20 variable factors may be involved [96].

B. Functional Requirements of Proteins as Food Ingredients

Morr [97] has listed a number of requirements related to different properties of proteins as food ingredients, as indicated in Table 2. The major functional properties of food proteins that influence their choice in specific applications are emulsification, aeration/foam formation, gelation, solubility/water binding, dispersibiliy, film formation (cohesion/adhesion), heat stability, and acid stability.

The first three have been related to surface properties in the first part of this chapter. Also, in the broadest sense, the remaining properties can also be associated with protein surface properties. In most food systems, multiple functional properties of proteins are required

Table 2 Functional Requirements of Food
Protein Ingredients

Property	Functional attribute
Sensory	Flavor, odor, texture, color
Visual	Opacity, turbidity, color
Hydration	Solubility, dispersibility, gelation, viscosity
Surfactant	Emulsion, foaming, whipping, baking
Textural	Viscosity, adhesion, aggregation, gelation
Rheological	Aggregation, gelation, viscosity, extrudability
Other	Comparability with other ingredients and with processing conditions

to yield the desired characteristics, even though a particular functional property can be more important than others. The characteristic functional property associated with different food types is given in the following examples:

Emulsification: salad dressing, non-dairy coffee whiteners, instant breakfasts and infant formulas

Aeration/whipping/foaming: hot cocoa mix, whipped topping, frozen desserts and cakes

Gelation: Breads, restructured meats, puddings, gelatin desserts

Film formation/cohesion/adhesion: batters, glazes, breakfast bars, edible packaging films

For most food systems, more than one functional property is required. In the formulation of frying batters for fish and poultry for example, emulsification, adhesion, cohesion, and gelation are all required. Foaming during the frying process may be required for some batter applications. Another example is in the manufacture of a high-ratio cake, where emulsification, foaming, cohesion and gelation are all required. Functional properties required for different applications has been reviewed by Morr [98], de Wit [89], and Kinsella and Whitehead [40]. An adaptation and extension of the multiple func-

Table 3 Food Protein Requirements for Application in Different Food Products

Food Product	Required functions for all products	Additional functions for some products
Beverages	Solubility, colloidal stability	Acid stability, emulsifying, water binding
Bakery	Solubility, emulsifying, gelation	Foaming, foam stability, water binding, gluten modification
Confectionery	Foaming, solubility	Emulsifying, gelation
Frozen desserts	Emulsifying, foaming, dispersibility	Solubility, water binding, fat mimetic
Imitation dairy	Emulsifying, colloidal stability	Solubility, foaming, foam stability
Infant formula	Nutrition, solubility, emulsification, colloidal stability to heat	Mimic human milk composition
Reformed meat	Emulsification, water binding, emulsifying	Salt solubility, low viscosity in solution, gelation, fat mimetic
Retort stable sauces	Emulsifying, colloidal stability to heat	Water binding, viscosity building
Salad dressing	Emulsifying, acid stability	Fat mimetic

tional properties required is given in Table 3. Where multiple processing steps are involved, the major functional property involved at each step in the process may be different. In fact, modification of the protein at one step in the process may be essential for the protein to achieve the desired functionality at the next step in the process.

C. Factors Influencing Protein Performance in Food Products

Both intrinsic and extrinsic factors influence the performance of food proteins. The intrinsic properties are those inherent to the protein, but they can be variable between different lots produced by the same protein manufacturer. Extrinsic factors depend upon the product being produced, the processes used, and the equipment employed. In some instances the protein-defined characteristics of a given food processed by the same procedure will be different because of differences in the equipment used in different plants.

The major factors that influence the success of a protein as a food ingredient include protein source, processing history of the protein, food formulation, environment, and food processing methods and equipment. In many cases the functionality of the protein may be modified during its preparation for better or for worse.

1. Protein Source

Differences in the major food proteins that influence their application as food ingredients reflect the intrinsic characteristics of those proteins. Tabulation of some of these differences that impact on the type of food in which they may be used is shown in Table 4. In addition to general functional properties, proteins to be used as food ingredients must also be easy to use, neither impart flavor nor change the

Table 4 Functional Characteristics of Some Common Food Proteins

Protein	Emulsifying property	Whipping property	Gelation property	Film formation	Environmental stability
Egg white	low	high	high	medium	unstable to heat
Egg yolk	high	low	medium	low	unstable to heat
Caseinate	high	medium	low	high	heat stable, unstable to acid
Whey protein	medium	low to high	low to high	medium	acid stable, unstable to heat
Soy isolate	medium to high	low to medium	medium	medium to high	unstable to heat and acid
Fish protein	medium	low	high to medium	low to medium	unstable to heat

flavor of the food, and be applicable without need for the processor to modify existing formulations or make a capital expenditure. Ease of use covers a number of attributes, which include

Packages that are easy to handle and open
Product that does not cake on storage
Product that retains its functional characteristics during its promised shelf life
Product that has a high bulk density and does not create dust when used

Flavor properties of proteins are critical in food applications. Frequently, products with the same or equal functionality will not be chosen for use by food manufacturers because of their effect on flavor. This includes not only off-flavors associated with the protein but also the binding of added flavors by the protein, thus reducing the impact of the added flavor. Flavor binding by food proteins has been reviewed by Jasinski and Kilara [99] and Kinsella [100].

Prior discussion in this book has focused on purified proteins. However, food protein ingredients are not generally simple proteins but rather mixtures of components ranging in total protein content from 35 to 95%. Each protein in the ingredient contributes its own functional characteristics, and the overall function of the protein ingredient in the food is influenced by the sum of the component proteins and their physicochemical form. For example, casein is comprised of three main proteins, beta, $alpha_s$, and kappa. In the native state casein occurs in micellar form with particle weights in excess of one million daltons, whereas in sodium caseinate the monomeric form is prevalent with molecular weights of approximately 20,000 daltons. As a result, the emulsifying properties of sodium caseinate are much great than those of micellar casein. This is the reason that sodium caseinate is used in preference to micellar casein or calcium caseinate in products such as coffee whiteners and whipped toppings [88]. In bread manufacture, glutenin controls mixing time, whereas gliadin influences loaf volume [101]. Thus variations in the ratios of these two proteins in different wheat varieties has an influence on the baking properties of wheat flour.

In addition to containing a combination of individual proteins, protein ingredients generally include lipid, carbohydrate, and minerals to varying degrees, which can modify the performance of the product in food applications. Composition may vary as a function of the breed/variety, the season of the year, and the process used to obtain the protein ingredient. Commercial suppliers of proteins provide a range of protein products that have been prepared for different applications. The proprietary product Simplesse represents a process-modified protein where microparticulation is claimed to provide textural attributes useful in fat simulation. However, frequently, the differences are related to variations in chemical composition.

2. Protein Processing

As stated previously, the functional performance of food proteins as ingredients is dictated by surface characteristics, including hydrophobicity, electrostatic, and steric parameters [100]. All of these properties can be modified by the processing utilized to provide a concentrated food protein ingredient. Although the process modification that influences the functionality of a food protein will be specific to the type of protein processed, there are some general characteristics that apply equally to all common food proteins. These include

Method of handling the raw material, including time and temperature of holding the material, pretreatment, and the amount of shear used in transporting the product
Method of concentration of the protein
pH during processing
Ionic strength during processing
Method of drying
Packaging materials and conditions of storage

The effect of processing on protein functionality has been reviewed extensively [102,103,104,105,106,107]. In many cases the exact significance of all of the steps in the process of making protein ingredients is not well understood. In this respect, research on the effect of processing on whey protein functionality has been extensive. Hobman [102] indicated that this protein ingredient is especially sensitive to alterations in functionality through processing and that each step in the process can influence the functional quality. However, the exact physicochemical modifications of the individual proteins is poorly understood. Very minor variations in a process have been demonstrated to alter the functional performance of whey proteins. Chen [107] reported that changing the pasteurization temperature of milk by as little as 2°C could alter the functional properties of 75% whey protein concentrate subsequently made from that milk. Such effects can be expected for other food protein ingredients as more detailed investigations are made of the effect of processing on performance.

Proteins can be concentrated by isoelectric precipitation, heat precipitation, alcohol precipitation, ultrafiltration, microfiltration, and freeze concentration. Numerous references report differences in function in relation to the method of concentration of the protein. Whey protein produced by acid–heat precipitation (lactalbumin) is essentially insoluble but shows high water binding capacity and for this reason is the whey protein of choice for baking and for formulation of cereal products [89]. The use of ultrafiltration instead of acid precipitation of soy protein provides products with quite different characteristics [108]. However, caseinate produced via acid precipitation is nearly as suitable for the production of imitation cheese as is rennet casein produced by the action of chymosin on skim milk [88].

High heat treatment of proteins during their preparation results in a reduction in solubility, an increase in water holding capacity, and variable effects on other properties [81]. There is general agreement that very high heat treatment during the preparation of proteins has an adverse effect on functionality. However, mild heat treatment has been beneficial to certain applications of proteins in some food. Heat effects have been more widely studied in respect to whey proteins than for other common proteins used as food ingredients. Zadow and Harper [109] found that partial denaturation of WPC retentate improved the loaf volume and texture of white bread, whereas high levels of denaturation had an adverse effect on bread quality. Moderate denaturation of whey protein isolate has been found to increase the polymeric species of protein and to improve foaming properties. Huang [110] used microwave heating at 185°F in place of standard pasteurization to extend the shelf life of egg whites, without adversely affecting their subsequent application.

Drying methods of protein ingredients can substantially affect their subsequent application. Knorr [111] reported that the method of drying significantly affects solubility and water binding properties of plant protein concentrates. There were marked effects of interactions between the method of drying, pH at drying, and drying temperature. Mangino et al. [81] found that spray drying WPC retentates increased surface hydrophobicity and decreased alkane binding values of WPC, which would alter their functionality in respect to food applications, especially as related to emulsification. Schmidt et al. [112,113] reported that drying temperature influenced emulsification and gelation.

3. Food Formulation and Environment

For any specific food product that uses a protein as a functional ingredient, the general formulation and the processing environment dictate how the protein will function. Most formulated protein-based foods include, in varying concentrations, protein, lipid, simple carbohydrates as sweeteners, complex carbohydrates as stabilizers, small molecular weight emulsifiers, and minerals (salts).

Each of the components in the food system is certain to modify the surface properties of the protein and its functional performance [88,89,90]. These modifications can occur through

Changes in protein configuration
Changes in protein surface charges
Interactions between ingredients
Competition at oil or air interfaces

4. Food Processing

Processing is known to affect the functional performance of food protein ingredients, but relatively little information is available in the published literature concerning the specific effects of processing of food on the functional effectiveness of food protein ingredients [87,88,89,114]. In general heat and shear processes are known to influence the success of food proteins in food applications. A protein product that works well for one manufacturer of a given food product may not perform for another. This is thought to be related to differences in equipment used by different food manufacturers as well as small changes in heat and shear effects. This very complex area requires much further investigation [115].

D. Selected Food Applications

The best way to illustrate the factors affecting the application of proteins in food systems is to describe our current knowledge for selected food products that relate to surface functional properties. For this purpose, three types of products have been selected, all of which require emulsification as one functional property. The first type, coffee whitener, requires only emulsification of the major functional protein properties. The second type, whipping cream/whipped toppings, requires both emulsification and whipping. The third type, cakes, requires whipping and gelation for all types, while emulsification is required for cakes containing shortening.

1. Emulsification (Coffee Whitener)

Coffee whiteners have the typical formulation

Oil/fat	5–10%
Protein	0.5–2.0%
Carbohydrate	5%–10%
Emulsifiers	0.1%
Phosphates	0.1%
Gums	0.05–0.1%

All of the ingredients are interactive in the coffee whitener system. Pearce and Harper [116], in a study of a coffee whitener made with soy proteins, found that the type and ratio of emulsifiers influences the stability of the coffee whitener, the ability to lighten coffee, and the stability of the whitener in the coffee. Harper [114], using a model coffee whitener system, established that three characteristics of the coffee whitener emulsions could be influenced independently by the protein used in the system. These were fat globule size, hydrated bound protein layer, and surface performance of the emulsification on storage. Each of the parameters was influenced by the ingredients in the formulation. Whey proteins and sodium caseinate gave different characteristics to the coffee whitener. Using an emulsifier with a high HLB value resulted in coffee whitener with smaller fat globules but poor storage stability when made with sodium caseinate. Different types of whey proteins gave coffee whiteners with different characteristics.

Although sodium caseinate is the most common protein used in coffee whiteners, a range of oil seed proteins have also been utilized. One limitation of these products has been the flavor that they contribute to the whitener.

2. Emulsification and Whipping (UHT Whipping Cream)

Whipped toppings can be divided into four classes:

Whipped cream
UHT whipped cream
Whipped topping
Aerosol whipped cream and topping

In all cases the final foam structure is stabilized by agglomeration of the fat globules that occurs during whipping.

The mechanisms involved in this agglomeration are different in the four types of whipped products, and the fat globule membrane appears to have a different structure at the time of whipping. The propellant of choice in aerosol products is nitrous oxide. The solubility of this gas in the fat phase of the products is thought to provide the primary basis for disruption of the fat globule membrane and resultant fat globule agglomeration.

UHT whipping cream is intermediate between a standard whipping cream and a nondairy whipped topping. In addition to cream, the product will also contain emulsifiers and hydrocolloid stabilizers. Homogenization, which is deleterious to standard whipping cream, is essential to UHT whipping cream.

Homogenization or UHT pasteurization of whipping cream results in a stronger fat globule membrane, which resists agglomeration upon cooling. Towler [117] found that the addition of 0.1% emulsifier reduced whipping time and increased overrun in UHT whipping creams. Homogenization is essential to provide a stable emulsion. The role of the emulsifier is considered to be similar to that in ice cream, where the emulsifier results in desorption of the protein during cool aging of the cream. This desorption weakens the membrane around the fat globule and permits fat agglomeration during whipping. Bruhn and Bruhn [118] found the processing aids (emulsifiers and gums) nearly doubled the overrun compared to standard cream.

Whipped toppings, which serve the same purpose as whipping cream, substitute solid fats (such as coconut) for milk fat and incorporate functional additives. A typical formulation is

Coconut oil	30%
Protein	2%
Sugar	5%
Corn syrup	5%
Gum stabilizer	0.1%
Emulsifier	0.2%

The ingredients are mixed, pasteurized, homogenized, aged under refrigeration, and whipped. Liao and Mangino [9], using WPC, found that increasing surface hydrophobicity increased overrun in a whipped topping. For a soybean-based whipped topping, Chow [119] found that a blend of hydrophilic and hydrophobic emulsifiers gave better overrun and stiffness than did either emulsifier on its own. Min and Thomas [120] used surface response methodology and reported that each ingredient mainly reacted independently in contributing to overrun and firmness of a caseinate-based topping. Different combinations of protein, stabilizer, and corn syrup could be used to give the same properties to the topping.

3. Cakes

All cakes contain flour, sugar, egg (or egg substitute), and leavening and leavening agents. Some but not all cakes contain a shortening. The ratio of flour, sugar, and shortening will vary with the type of cake. The final structure of cakes is dependent upon starch gelatinization during early baking and the denaturation of wheat protein to "set" the final structure. Unlike most breads, cakes have a high sugar content, which increases the denaturation temperature of the wheat flour. For this reason, proteins with lower gelation temperature than gluten are added to cakes to provide the final structure. In cakes containing lipids, eggs serve multiple purposes:

Encapsulation of fat droplets (emulsification)

Stabilization of air micelles in the batter

Providing final heat set through coagulation of protein at the end of the baking process

Cakes illustrate a class of food products in which the proteins influence more than one characteristic of the product. While starch is a main structural entity, the proteins have been shown to control cake volume and tenderness and influence crumb dryness and cracking, flavor, structure, and crumb and crust color and conformation.

In recent years whey protein concentrates have been tried as replacements of egg proteins in cakes. According to de Wit [89], low protein WPC could not replace whole eggs in Madeira cakes because such replacement resulted in collapse during the later stages of the baking process. This problem was solved, in part by lowering the temperature of gelation of the WPC to near that of egg white, or by previously emulsifying the shortening with the whey protein before making the batter. Harper [115] compared the use of 75% WPC at the same protein level in four types of cake. The effects on volume, tenderness, and crumb character are given in Table 5. The mechanisms to explain the differences shown in this table are not understood at this time. Devilbliss et al. [50] showed that the failure of WPC to perform in angel food cake could be related to the fact that egg white protein was denatured by shearing forces during batter preparation, whereas whey protein concentrate did not. In cakes containing shortening, Mann et al. [121] found that a whey protein–

Table 5 Effect of WPC on the Characteristics of Different Types of Cakes

Cake type	Volume	Tenderness	Crumb character
White cake	good	excellent	brown, speckled
Pound cake	good	good	brown, cracked
Sponge cake	fair	good	sticky, brown
Anglefood cake[a]	unsatisfactory		

[a]Cake gave good volume at end of baking and collapsed after being removed from the oven.

Table 6 Guidelines for the Use of Selected Food Proteins

Protein	Functional property	Examples
WPC	Emulsification	Cake mixes, sauces, toppings, bakery, frozen desserts, salad dressings, creaming stability
	Foaming	Whipped toppings, frozen deserts, confectionery, bakery
	Thermal gelation	Restructured meat products, fat mimetics, low-fat bakery
	Acid stability	Beverages, sauces, salad dressings
Caseinates	Emulsification	Coalescence stability, high-fat frozen desserts
	Thermal stability	Sterile nutritional products, coffee whiteners, retortable products
	Melting properties	Imitation cheese products, cheese spreads
	Turbidity	Beverages, some soups and sauces, coffee whiteners
Egg white	Foaming	Cakes, meringues, confectionery, bread and other bakery
	Thermal gelation	Restructured meat products, fat mimetics, angel food cakes, bakery
Egg yolks	Emulsification	Salad dressings
	Gelation	Some bakery, custards

carboxymethyl cellulose formulation could be used successfully to replace whole egg in cakes. The CMC was thought to raise the viscosity sufficiently so that the poorer gelling property of the WPC could be overcome.

These few examples demonstrate the difficulty faced by food processors in the selection of functional ingredients. There are no simple rules that can be applied to the selection of a given protein ingredient for a given functionality that are applicable to all or even most products. The fact that there are many acceptable products produced indicates that processors have developed empirical selection criteria that tend to be product specific. As more is learned regarding the fundamental properties that are necessary for selected functionalities in selected food products, and as more information is gained regarding the properties that contribute to the functionalities of isolated proteins in relatively simple applications, we can anticipate the development of better selection criteria. As more research becomes available regarding the complex interactions of proteins with themselves and with other food ingredients, and as more complex models are investigated, we will be able to provide the processor with a more comprehensive description of the surface properties essential to protein functionality. Increased research conducted at the interface between the collection of data from very simple mixtures and the formulation and processing of real food products is necessary to generate this information. While there are no simple rules that are applicable to all functional properties or to all products, it is possible to use the considerable research that has been done in the area of protein functionality to establish some guidelines for protein selection. Table 6 lists some of the functional properties of the major food proteins and gives examples of how they can be used.

REFERENCES

1. E. Tornberg and A.-M. Hermansson, *J. Food Sci. 42*: 468 (1977).
2. E. Tornberg, A. Olsson, and K. Persson (1990), in *Food Emulsions*, 2d ed. (K. Larson and S. T. Friberg, eds.), Marcel Dekker, New York, 1990, pp. 247–326.
3. P. Walstra, in *Food Emulsions and Foams* (E. Dickinson ed.), Royal Society of Chemistry, London, 1987, pp. 242–257.
4. M. E. Mangino, *J. Dairy Sci. 67*: 2711 (1984).
5. D. F. Darling and R. J. Birkett, in *Food Emulsions and Foams* (E. Dickinson, ed.), Royal Society of Chemistry, London, 1987, pp. 1–29.
6. P. J. Halling, *CRC Crit. Rev. Food Sci. Nutr. 15*: 155 (1981).
7. J. N. de Wit, G. Klarenbeek, and M. Adamse, *Neth. Milk Dairy J. 40*: 41 (1986).

8. J. E. Kinsella, *CRC Crit. Rev. Food Sci. Nutr. 7*: 219 (1976).
9. S. Y. Liao and M. E. Mangino, *J. Food Sci. 52*: 1033 (1987).
10. R. Peltonen-Shalaby and M. E. Mangino, *J. Food Sci. 51*: 103 (1986).
11. Mangino, M. E., in *Protein Functionality in Foods* (N. S. Hettiarachchy and G. R. Ziegler, eds.), Marcel Dekker, New York, 1994, pp. 147–179.
12. R. D. Waniska and J. E. Kinsella, *Food Hydrocolloids 2*: 59 (1988).
13. H. E. Swaisgood, in *Developments in Dairy Chemistry 1. Proteins* (P. F. Fox, ed.), Applied Science, London, 1982, pp. 1–59.
14. F. MacRitchie, *Adv. Protein Chem. 32*: 283 (1978).
15. C. Tanford, *Adv. Prot. Chem. 24*: 1 (1970).
16. D. E. Graham and M. C. Phillips, *J. Colloid Interface Sci. 70*: 403 (1979).
17. G. Stainsby, in *Functional Properties of Food Macromolecules* (J. R. Mitchell and D. A. Ledward, eds.), Elsevier, London, 1986, pp. 315–353.
18. S. H. Kim and J. E. Kinsella, *J. Food Sci. 52*: 128 (1987).
19. E. Tornberg, *J. Sci. Food Agr. 29*: 904 (1978).
20. J. A. Hunt and D. Dalgleish, *Food Hydrocolloids 8*: 175 (1994).
21. C. Tanford, *The Hydrophobic Effect: Formation of Micelles and Biological Membranes*, 2d ed., John Wiley, New York, 1980.
22. G. E. Petrowski, *Adv. Food Res. 22*: 110 (1976).
23. B. A. Bergenstål and P. M Claesson, in *Food Emulsions*, 2d ed. (K. Larson and S. T. Friberg, eds.), Marcel Dekker, New York, 1990, pp. 41–96.
24. S. E. Friberg, R. F. Goubran, and I. H. Hayai, in *Food Emulsions*, 2d ed. (K. Larson and S. T. Friberg, eds.), Marcel Dekker, New York, 1990, pp. 1–40.
25. P. M. T. Hansen, *Prog. Food Nutrition Sci. 6*: 127 (1982).
26. S. K. Sharma, *Food Technol. 35*: 59 (1981).
27. K. L. Fligner, M. A. Fligner, and M. E. Mangino, *Food Hydrocolloids 4*: 95 (1990).
28. S. K. Sharma and D. G. Dalgleish, *J. Agric. Food Chem. 41*: 1407 (1993).
29. E. Dickinson, S. E. Rolfe, and D. G. Dalglish, *Food Hydrocolloids. 3*: 193 (1989).
30. J.-L. Courthaudon, E. Dickinson, and W. W. Christie, *J. Agric. Food Chem. 39*: 1365 (1991).
31. E. Dickinson and S. Tanai, *Food Hydrocolloids 6*: 163 (1992).
32. E. Dickinson and G. Iveson, *Food Hydrocolloids 6*: 533 (1993).
33. K. Yamauchi, M. Shimizu, and T. Kamiya, *J. Food Sci. 45*: 1237 (1980).

34. K. L. Fligner, M. A. Fligner, and M. E. Mangino, *Food Hydrocolloids. 5*: 269 (1991).

35. R. Ponnampalam, G. Goulet, J. Amiot, B. Chamberland, and G. J. Grisson, *Food Chemistry 29*: 109 (1988).

36. R. Jimenez-Flores and T. Richardson, *J. Dairy Sci. 71*: 2640 (1988).

37. J. Haque, presented by Z. Haque, American Dairy Science Association Annual Meeting, Columbus, Ohio, 1992.

38. J. B. German and L. Phillips, in *Protein Functionality in Foods* (Hettiarachchy and G. R. Ziegler, eds.), Marcel Dekker, New York, 1994, pp. 181–208.

39. M. E. Mangino, in *Whey and Lactose Processing* (J. G. Zadow, ed.), Elsevier, Amsterdam, 1992, pp. 231–270.

40. J. E. Kinsella and Whitehead, *Adv. Food Nutrition Res. 33*: 403 (1989).

41. R. Nakumara and Y. Sato, *Agric. Biol. Chem. 31*: 530 (1964).

42. M. C. Phillips, *Food Technol. 35*: 50 (1981).

43. K. L. Fligner and M. E. Mangino, in *Food Proteins: Stability and Interactions* (N. Parris, ed.), ACS Press, Washington, D.C., 1991, pp. 1–12.

44. S. Poole and J. Fry, in *Developments in Food Proteins 5* (B. Hudson, ed.), Elsevier Applied Science, New York, p. 257.

45. A. Prins, in *Food Emulsions and Foams* (E. Dickinson and G. Stainsby, eds.), Elsevier Applied Science, New York, 1988, pp. 91–122.

46. E. Tornberg, *J. Food Sci. 43*: 599 (1978).

47. L. G. Phillips, W. Schulman, and J. E. Kinsella, *J. Food Sci. 55*: 588 (1990).

48. L. G. Phillips. Relationship between structural, interfacial and foaming properties of β-lactoglobulin. Ph.D. dissertation, Cornell University, Ithaca, New York, 1992.

49. S. Damodaran, in *Protein Functionality in Foods* (N. S. Hettiarachchy and G. R. Ziegler, eds.), Marcel Dekker, New York, 1994, pp. 1–37.

50. E. D. Devilbliss, V. H. Holsinger, L. P. Posati, and M. J. Pallansch, *Food Technol. 28*: 40 (1974).

51. S. H. Richert, C. V. Morr, and C. M. Cooney, *J. Food Sci. 39*: 42 (1974).

52. A. Kato, H. R. Ibrahim, H. Watanabe, K. Honma, and K. Kobayashi, *J. Agric. Food Chem. 37*: 433 (1989).

53. H. R. Ibrahim, K. Kobayashi, and A. Kato, *Biosci. Biotech. Biochem. 57*: 1549 (1993).

54. T. O. R. Haggett, *N.Z. J. Dairy Sci. Technol. 11*: 275 (1976).

55. L. G. Phillips, S. E. Hawks, and D. M. Barbano, presented by L. G. Phillips, Institute of Food Technologists Annual Meeting, Atlanta, Georgia, 1994.
56. M. S. B. Joseph and M. E. Mangino, *Australian J. Dairy Technol. 5*: 6 (1988).
57. M. S. B. Joseph and M. E. Mangino, *Australian J. Dairy Technol. 5*: 9 (1988).
58. J. B. German, T. O'Neill, and J. E. Kinsella, *J. Am. Oil Chem. Soc. 62*: 1358 (1985).
59. N. Kella, T. Yang, and J. E. Kinsella, *J. Agric. Food Chem. 37*: 1203 (1989).
60. M.-A. Yu and S. Damodaran, *J. Agric. Food Chem. 39*: 1555 (1991).
61. L. G. Phillips, S. T. Yang, W. Schulman, and J. E. Kinsella, *J. Food Sci. 54*: 743 (1989).
62. S. Poole, S. West, and C. Walters, *J. Sci. Food Agric. 35*: 701 (1984).
63. G. R. Ziegler and E. A. Foegeding, *Adv. Food Res. 34*: 203 (1990).
64. P. J. Flory, *J. Amer. Chem. Soci. 63*: 3083 (1941).
65. A. M. Hermansson. in *Food Texture and Rheology* (P. Sherman, ed.), Academic Press, New York, 1979, p. 266.
66. R. H. Schmidt, in *Protein Functionality in Foods* (J. R. Cherry ed.), American Chemical Society, Washington, D.C., 1981, pp. 131–147.
67. K. Shimada and J. C. Cheftel, *J. Agric. Food Chem. 36*: 1018 (1988).
68. K. Shimada and J. C. Cheftel, *J. Agric. Food Chem. 37*: 161 (1989).
69. C. E. Lupano, E. Dumay, and J. E. Cheftel, *Internat. J. Food Sci. Technol. 27*: 615 (1992).
70. S. Barbut and E. A. Foegeding, *J. Food Sci. 58*: 867 (1993).
71. R. H. Schmidt, L. B. Illingworth, J. C. Deng, and A. J. Cornell, *J. Agric. Food Chem. 27*: 529 (1979).
72. F. Zirbel and J. E. Kinsella, *Milchwissenschaft 43*: 691 (1988).
73. K. R. Langley and M. J. Green, *J. Dairy Res. 56*: 275 (1989).
74. Y.-A. Kim, G. W. Chism, and M. E. Mangino, *J. Food Sci. 52*: 124 (1987).
75. J. A. Dunkerley and J. F. Hayes, *N.Z. J. Dairy Sci. Technol. 15*: 191 (1980).
76. M. E. Mangino, J. H. Kim, J. A. Dunkerley, and J. G. Zadow, *Food Hydrocolloids 1*: 227 (1987).
77. A. L. Kohnhorst and M. E. Mangino, *J. Food Sci. 50*: 1403 (1985).
78. A. Kato and S. Nakai, *Biochim. Biophys. Acta 624*: 3 (1980).
79. L. P. Voutsinas, E. Cheung, and S. Nakai, *J. Food Sci. 48*: 26 (1983).

80. L. P. Voutsinas, S. Nakai, and V. R. Harwalker, *Can. Inst. Food Sci. Technol. J. 16*: 185 (1983).

81. M. E. Mangino, S. Y. Liao, W. J. Harper, C. V. Morr, and J. G. Zadow, *J. Food Sci. 52*: 1522 (1987).

82. D. M. Mulvihill and J. E. Kinsella, *J. Food Sci. 53*: 231 (1988).

83. G. W. Froning, in *Developments of Food Proteins 6* (B. F. J. Hudson, ed.), Applied Science Publishers, London, 1982, pp. 1–34.

84. T. Inomata, *New Food Industry 26*: 40 (1984).

85. D. Fukushima, *Food Reviews International 7*: 323 (1991).

86. A. Visser and A. Thomas, *Food Reviews International 3*: 1 (1987).

87. E. W. Evans, in *Developments of Food Proteins 1* (B. F. J. Hudson, ed.), Applied Science Publishers, London, 1982, pp. 131–169.

88. C. R. Soutward, in *Developments in Dairy Chemistry 4* (P. F. Fox, ed.), Elsevier Applied Science, London, pp. 173–245.

89. J. N. de Wit, in *Developments in Dairy Chemistry 4* (P. F. Fox, ed.), Elsevier Applied Science, London, pp. 323–376.

90. J. D. Schofield and M. R. Booth, in *Developments of Food Proteins 2* (B. F. J. Hudson, ed.), Applied Science Publishers, London, 1982, pp. 1–66.

91. I. M. Mackie, in *Developments of Food Proteins 2* (B. F. J. Hudson, ed.), Applied Science Publishers, London, 1982, pp. 215–265.

92. W. J. Harper, *J. Dairy Sci. 67*: 2745 (1984).

93. A. T. Paulson, M. A. Tung, M. R. Garland, and S. Nakai, *Can. Inst. Food Sci. Technol. J. 17*: 202 (1984).

94. G. E. Arteaga and S. Nakai, *J. Food Sci. 58*: 1152 (1993).

95. G. E. Arteaga, E. Li-Chan, S. Nakai, S. Cofrades, and F. Jiminex-Colmenero, *J. Food Sci. 58*: 656 (1993).

96. J. Dou, S. Toma, and S. Naka, *Food Research International 26*: 27 (1993).

97. C. V. Morr, *N.Z. J. Dairy Sci. Technol. 14*: 195. 1979.

98. C. V. Morr, in *Developments in Food Chemistry 1* (P. F. Fox, ed.), Elsevier Applied Science, London, pp. 375–399.

99. E. Jasinski and A. Kilara, *Milchwissenschaft 40*(10): 596 (1985).

100. J. E. Kinsella, in *ACS Symposium Series 206*, American Chemical Society, pp. 301–326 (1982).

101. F. R. Huebner, *Baker's Digest 51*: 25. (1977).

102. P. Hobman, in *Whey and Lactose Processing* (J. G. Zadow, ed.), Elsevier, Amsterdam, 1992, pp. 195–220.

103. J. W. Finley, in *Protein Quality and the Effects on Processing* (R. D. Phillips and J. W. Finley, eds.), Elsevier Science Publishers, London, pp. 1–7.

104. S. Nakai and E. Li-Chan, *Protein Quality and the Effects on Processing* (R. D. Phillips and J. W. Finley, eds.), Elsevier Science Publishers, London, pp. 125–144.
105. R. D. Phillips, *Protein Quality and the Effects on Processing* (R. D. Phillips and J. W. Finley, eds.), Elsevier Science Publishers, London, pp. 219–246.
106. C. V. Morr, in *Proceedings of the Dairy Products Technical Conference*, April 25–26, Chicago, Illinois, 1992.
107. S.-I. Chen, Ph.D. dissertation, Ohio State University, 1993, pp. 1–129.
108. D. Knorr, *J. Food Process Engineering 5*: 215 (1982).
109. W. J. Harper and J. G. Zadow, *N.Z. J. Dairy Sci. Technol. 19*: 229 (1984).
110. F. Huang, U.S. patent 222,658 (1989).
111. D. Knorr, *J. Food Process Engineering 5*: 215 (1982).
112. R. H. Schmidt, V. S. Packard, and H. A. Morris, *J. Dairy Sci. 67*: 2733 (1984).
113. R. H. Schmidt, V. S. Packard, and H. A. Morris, *J. Food Protection 49*: 192 (1986).
114. W. J. Harper, in *Proceedings CRD/ADPI Whey Protein Workshop*, Madison, Wisconsin, 1991, pp. 55–78.
115. W. J. Harper, in *Proceedings CRD/ADPI Whey Protein Workshop*, Madison, Wisconsin, 1991, pp. 136–160.
116. R. J. Pearce and W. J. Harper, *J. Food Sci. 47*: 680 (1982).
117. C. Towler, *N.Z. J. Dairy Sci. Technol. 23*: 109 (1988).
118. C. M. Bruhn and J. C. Bruhn, *J. Dairy Sci. 71*: 857 (1988).
119. E. T. S. Chow, *J. Food Sci. 53*: 1765 (1988).
120. D. B. Min and E. L. Thomas, *J. Food Sci. 45*: 346 (1980).
121. R. S. Mann, B. N. Mathuv, and M. R. Srinivsan, *J. Food. Sci. Technol. 12*: 214 (1977).

Index

RETURN TO ➡

CHEMISTRY LIBRARY
100 Hildebrand Hall • 642-3753

LOAN PERIOD 1	2	3
	1 MONTH	
4	5	6

ALL BOOKS MAY BE RECALLED AFTER 7 DAYS
Renewable by telephone

DUE AS STAMPED BELOW

FORM NO. DD5

UNIVERSITY OF CALIFORNIA, BERKELEY
BERKELEY, CA 94720-6000